Infrared Spectroscopy - Perspectives and Applications

Edited by Marwa El-Azazy, Khalid Al-Saad and Ahmed S. El-Shafie

Published in London, United Kingdom

Infrared Spectroscopy - Perspectives and Applications
http://dx.doi.org/10.5772/intechopen.100794
Edited by Marwa El-Azazy, Khalid Al-Saad and Ahmed S. El-Shafie

Contributors
Yun-Hsuan Chen, Mohamad Sawan, Sonia Nieto-Ortega, Ángela Melado-Herreros, Idoia Olabarrieta, Giuseppe Foti, Rebeca Lara, Prasanta Das, Aulia M. T. Nasution, Hery Suyanto, Maggy Sikulu-Lord, Rafael Maciel-de-Freitas, Daniel T. Hallinan Jr, Cecilia Carlota Barrera-Ortega, America Rosalba Vázquez Olmos, Roberto Isaac Sato Berrú, Pineda Dominguez Karla Itzel, Nabeel Sabeeh Othman, Marwa El-Azazy, Ahmed S. El-Shafie, Khalid Al-Saad

First published in London, United Kingdom, 2023 by IntechOpen
IntechOpen is the global imprint of INTECHOPEN LIMITED, registered in England and Wales, registration number: 11086078, 5 Princes Gate Court, London, SW7 2QJ, United Kingdom

British Library Cataloguing-in-Publication Data
A catalogue record for this book is available from the British Library

Additional hard and PDF copies can be obtained from orders@intechopen.com

Infrared Spectroscopy - Perspectives and Applications
Edited by Marwa El-Azazy, Khalid Al-Saad and Ahmed S. El-Shafie
p. cm.
Print ISBN 978-1-80356-281-0
Online ISBN 978-1-80356-282-7
eBook (PDF) ISBN 978-1-80356-283-4

For EU product safety concerns:
IN TECH d.o.o., Prolaz Marije Krucifikse Kozulić 3, 51000 Rijeka, Croatia,
info@intechopen.com or visit our website at intechopen.com.

Meet the editors

Dr. Marwa El-Azazy is an analytical chemist, experienced educator, and researcher with more than 20 years of teaching experience at several institutions. Dr. Marwa's main research interests include the construction of sensors (using microfluidic platforms for point-of-care-testing of drugs and bio-analytes and ion-selective electrodes), chemometrics and analytical method development, spectroscopic analyses of drugs and pharmaceuticals, synthesis and characterization of nanomaterials, and development of green chemistry approaches for wastewater treatment. Dr. Marwa has published more than forty-five refereed papers in prestigious international journals, several conference presentations, two book chapters, and several research grants. She serves as a reviewer for a variety of international journals.

Dr. Ahmed El-Shafie is a chemist with more than 10 years of experience in teaching and research. His research is mainly in inorganic and analytical chemistry, including spectrophotometric and fluorometric analysis of different drugs using experimental design. He is also interested in the synthesis of nano-sorbents and their application in removing heavy metals, dyes, and pharmaceutically active materials from wastewater. Dr. El-Shafie has published sixteen refereed papers in international peer-reviewed journals.

Prof. Khalid Al-Saad is an analytical chemist, experienced educator, and researcher with more than 20 years of teaching experience at Qatar University. His main research interests include applications and methodology development of spectroscopy, chromatography, and mass spectrometry for environmental and biological analysis. He is also interested in wastewater treatment. Dr. Khalid has published more than forty refereed papers in international journals, several conference presentations, one book chapter, and several research grants.

Contents

Preface

Infrared spectroscopy (IR) is a potent spectroscopic technique that has seen considerable developments over the past few decades. With a variety of applications in different fields, IR spectroscopy is becoming an indispensable analytical technique. Its ease of operation, its non-destructive nature, and its capability to elucidate structures of both organic and inorganic materials make IR exceptionally attractive for numerous applications. The typical applications of IR spectroscopy include the identification of functional groups, anomeric carbon structures, crystal allomorphs, and more. Because some structures are so similar, the qualitative characterization of these compounds using IR spectroscopy is a challenging task. Furthermore, the overlap of bands in spectra might obscure relevant information. As a result, qualitatively analyzing different compounds using only one characteristic IR peak may be purely empirical and inaccurate.

This book introduces the basic principles of IR spectroscopy with a focus on recent advances and applications. It includes ten chapters that examine the applications of IR spectroscopy in different disciplines. Chapter 1 by Dr. El-Azazy et al. discusses the basic principles of IR spectroscopy, the most common assignments of the different functionalities required by researchers, and the implementation of IR spectroscopy in different applications. Chapter 2 by Dr. D.T. Hallinan Jr. presents a tutorial on the attenuated total reflectance (ATR) mode of Fourier-transform infrared (FT-IR) spectroscopy and how it can be used to measure transport through polymer membranes. In addition to covering the experimental setup and time-resolved data processing, this chapter presents fundamental equations for analyzing data to obtain diffusion coefficients. Chapter 3 by Dr. Prasanta Das investigates the instrumentation of gas phase FT-IR and its recent advancements and applications. The chapter focuses on the principle and data acquisition scheme of the repetitive mode measurement method of an FT-IR spectrometer. Chapter 4 by Prof. Dr. Nabeel Othman discusses both the qualitative and quantitative applications of IR spectroscopy. Chapter 5 by Sonia Nieto-Ortega et al. discusses the application of near IR spectroscopy (NIRS) in the food sector, especially for the fish value chain, combined with several chemometric techniques for quality, authenticity and safety applications. Chapter 6 by Dr. Chen et al. examines the applicability of functional NIRS (fNIRS) in monitoring brain activities to avoid stroke. Chapter 7 by Dr. Maggy et al. discusses the ability of NIR/MIR to provide fast and reliable estimates on subjects related to vector-borne diseases. Chapter 8 by Dr. El-Azazy et al. highlights the applications of FT-IR in the characterization of lignocellulosic biomasses and its impact on the depollution capacity for wastewater treatment. Chapter 9 by Aulia M T Nasution et al. discusses the employment of IR spectroscopy for detecting adulterants in food and traditional Indonesian herbal medicine. Finally, Chapter 10 by Dr. Barrera-Ortega et al. explores the application of Raman spectroscopy for dental enamel surface characterization.

I would like to thank my co-editors Prof. Khalid Al-Saad and Dr. Ahmed El-Shafie for their great efforts in reviewing and editing this book. I am indebted to all the authors for helping me to accomplish this project and who have contributed excellent chapters. All authors are experts in their fields and have tried their best to disclose the inherent concepts related to any of the mentioned applications. Their efforts to make this book comprehensive and informative are much appreciated.

I am most grateful to Author Service Manager Ms. Romina Rovan at IntechOpen for her efforts and support throughout the process of editing this book. Many thanks to the entire IntechOpen publishing team for making this book possible.

Dr. Marwa El-Azazy, Dr. Ahmed S. El-Shafie and Prof. Khalid Al-Saad
Department of Chemistry and Earth Sciences,
College of Arts and Sciences,
Qatar University,
Doha, Qatar

Chapter 1

Introductory Chapter: Infrared Spectroscopy - Principles and Applications

Marwa El-Azazy, Ahmed S. El-Shafie and Khalid Al-Saad

1. Spectroscopy

Spectroscopy is a term used for the study of the spectra generated as a result of interaction of electromagnetic radiation with matter. It is used for the detection and/or identification of atoms, molecules, functional groups, and nuclei based on the produced spectra following the interaction of matter with the radiation. When light strikes matters, it can physically be reflected, refracted, scattered, or transmitted. During transmittance through matter, light with specific energy may interact with molecules in different ways, depending on the energy (E, Joule) of lights, which is directly proportional to its frequency (υ, with the unit sec^{-1} or Hz) and inversely proportional to wavelength (λ) as according to the equations below-mentioned:

$$\text{Energy, } E = h.\upsilon \tag{1}$$

where h is Planck's constant ($6.6260755 \times 10^{-34}$ J·s)

$$\text{Frequency, } \upsilon = C/\lambda \tag{2}$$

where c is the speed of light (2.998×10^{8} m/s in vacuum).
From Eqs. (1) and (2),

$$E = h.C/\lambda \tag{3}$$

Expressing 1/λ as wavenumber ($\bar{\upsilon}$), Eq. (3) can be described as below:

$$E = h.C.\bar{\upsilon} \tag{4}$$

Based on the above relationships, the energy of light can be scaled on a sector of spectrum by the unit of wavelength (300–700 nm, in UV–visible spectrophotometry) or in wavenumbers (400–4000 cm^{-1}, for the mid-IR (MIR) region). Schematic description showing other electromagnetic regions of the spectrum, along with molecular processes rising from the interactions in each region, is illustrated in **Figure 1**. Matter may interact with light of microwave region causing the rotational movement of entire molecule. Light of particular energy in the IR region may cause the vibrational movement of bonds in the molecules, including stretching, bending,

IntechOpen

scissoring, rocking, wagging, and twisting, giving different patterns for different molecules and allowing the identification of different functional groups. UV–visible light may cause electron excitation. Ionization and bond breaking can happen in the X-ray region. The patterns of the UV and IR spectra extended on the x-axis work for the identification of different substances/molecules, while the intensities of the spectrum's peaks on the scale of y-axis correspond proportionally to the concentration of measured molecules, allowing the quantitative determination of molecules.

2. Infrared spectroscopy (IR)

Appearing at a longer wavelength compared to the visible region, as shown in **Figure 1**, the invisible IR radiation extends from the red edge of the visible region at 780 nm to 10^6 nm reaching the edge of microwave region. As shown in **Figure 1**, the IR region is divided into three main zones:

i. The far-IR (FIR, 400–10 cm^{-1}, 25–300 μm)

ii. The mid-IR (MIR, 4000–400 cm^{-1}, 2.5–25 μm)

iii. The near-IR (NIR, 14,000–4000 cm^{-1}, 0.7–2.5 μm)

Most bond stretches and other vibrations of the functional groups of the organic compounds occur at the MIR zone.

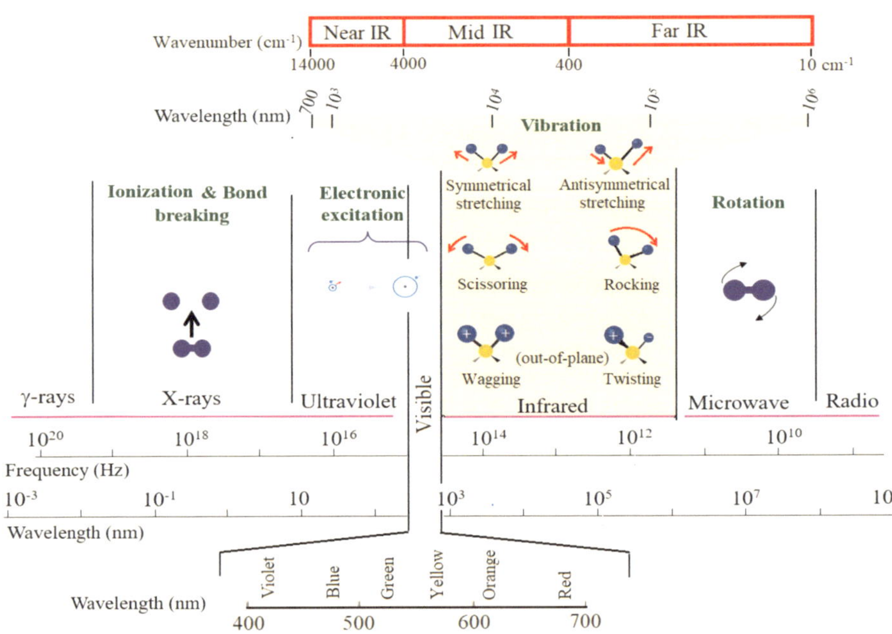

Figure 1.
Schematic description of electromagnetic radiation regions, showing representative molecular processes that occur in each region.

2.1 Assigning peaks in an IR spectrum

The MIR range can also be divided into two regions. The region below 1400 cm^{-1} is called "fingerprint region" giving peaks that are relatively difficult to assign, yet unique and specific for particular compounds. The region above 1400 cm^{-1} is called "diagnostic or functional group region," as it is often easier to pull specific details about the functional groups from this region. More specifically, regions can be divided into: 1) 2500–4000 cm^{-1} for Z-H single bond stretches (C—H, N—H, C—H); 2) 2000–2500 cm^{-1} for triple bond stretches (C≡C, C≡N); 3) 1500–2000 cm^{-1} for double bond stretches (C=C, C=O); 4) 1000–1500 cm^{-1} for single bond stretches (C—O); and 5) 1000 cm^{-1} for mostly bending vibrations. **Table 1**, below, summarizes the locations and nature of some functional group peaks in organic compounds.

2.2 Factors affecting peak location

In order to rationalize the different locations of bond stretches frequency in IR spectrum, we can think about bonds connecting two atoms, as springs connecting two objects. The factors affecting the stretches frequency of spring were described, in physics, by Hooke's law (Eq. (5)),

$$\upsilon = \frac{1}{2\pi c}\left[\frac{f(m_1 + m_2)}{m_1 m_2}\right]^{½}, \tag{5}$$

where f = force constant or bond strength and m = mass of atoms.

Based on this analogy, we can see that the locations of different stretches band are determined by two factors:

1. Bond strength: Stronger bond will need higher energy (higher wavenumber) to stretch (C=O > C—O & C≡C > C=C > C—C).

2. Mass of bonded atoms: If we are comparing stretching of same number of bonds but between two different atoms, the lighter the atoms connected, the greater the stretch frequency (H—Cl > H—Br > H—I).

2.3 Identification of organic compounds

IR spectroscopy was used to qualitatively analyze the organic compounds, such as identifying the polysaccharide type based on the presence of specific polysaccharides in a complex system [1]. Most of the absorption bands of the polysaccharides are present in the region between 1800 and 800 cm^{-1}. Canteri et al. (2019) revealed that the presence of several absorption bands in plant walls at 1075, 1440–1450, 1616, and 1740 cm^{-1} were distinguishing bands for pectin, and the absorption bands at 895, 1035–1041, and 1160–1163 cm^{-1} were used for cellulose [2]. Moreover, IR can also be used to identify substitution groups in the organic components, such as the presence of sulfate groups in the sulfated polysaccharides, which is confirmed by the presence of three characteristic bands at 1200–1270 cm^{-1}, 1010–1060 cm^{-1}, and 900–800 cm^{-1}, conforming to symmetric, asymmetric stretching of S=O, and stretching of C—O—S, respectively [3]. In addition, it can identify the anomeric carbon configurations by the presence of specific absorption bands in the anomeric region. The

Bond	Detailed type of bond	Detailed vibration type	Wavenumber (cm^{-1})	Intensity, appearance, and other peaks/pattern information for the identification of the functional groups
O—H	In water	Asymmetric stretch	3727	The stretching bands for O—H in water are weak and less broad compared to those for alcoholic OH
		Symmetric stretch	3607	
		In plane bending	1640	
	In alcohol	Stretching	3200–3650	Strong and broad due to hydrogen bonding
		Bending	1640	
	In carboxylic acid	Stretching	~3000	Strong, broad, and usually overlapped with C—H stretching.
N—H	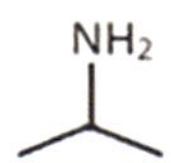	Primary amine N—H Stretching	3300 and 3500	Primary amine (R—NH_2) exhibits two N—H stretching bands at 3300 cm^{-1} and 3500 cm^{-1}, while secondary amine (R_2NH) exhibits one peak for N—H stretching at ~3300 cm^{-1}. Both types of amine give less broad and less intense bands than that for alcoholic O—H stretches bands.
	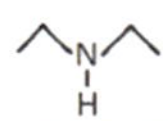	Secondary amine N—H Stretching	~3300	
C—H	In alkane	sp^3 stretching	< 3000	Stretch bands are relatively intense and quite narrow. C—H stretches occur between 2700 and 3300 cm^{-1}. Terminal alkyne having sp. C—H stretching will have very distinguished band at 3300 cm^{-1}. OH stretching band of carboxylic group often overlaps with the sp. C—H stretching bands (around 3200 cm^{-1}).
	In alkene	sp^2 stretching	> 3000	
	In alkyne	sp stretching		
		Bending	~1400	Bending bands are generally less pronounced than stretching bands. Terminal —CH_3 bending vibration raises just below 1400 cm^{-1}, while R_2CH_2 bending vibration will be raised just above 1400 cm^{-1}.
	In aldehydes 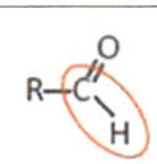	Fermi resonance in aldehyde	2820 and 2720	Doublet bands at 2820 and 2720 cm^{-1} occur due to fermi resonance in aldehyde.
C≡N		Stretch	2260–2220	The stretching band for C≡N is between 2260 and 2100 cm^{-1}. It is weak, similar to that for C≡C.

DOI: http://dx.doi.org/10.5772/intechopen.109139

Bond	Detailed type of bond	Detailed vibration type	Wavenumber (cm^{-1})	Intensity, appearance, and other peaks/pattern information for the identification of the functional groups
C≡C		Internal alkyne stretching	2260–2100	Weak to medium band occurs between 2260 and 2100 cm^{-1}, similar to that for C≡N. Both internal and terminal alkynes will have C≡C stretching band around 2200 cm^{-1}. However, way to distinguish the terminal alkyne is by the sp. C—H stretching that will be sharp and strong at ~3300 cm^{-1}. It is much narrower than the alcohol peak, which makes it recognizable. 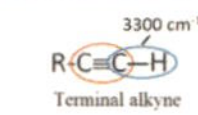
		Terminal alkyne stretching		
C=C	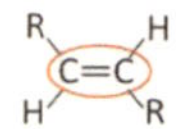	C=C stretching in alkanes	1680–1620	The stretching band region for C=C in alkene (1680–1620 cm^{-1}) is close to the C=O stretching band region. C=C stretching band is sharp, but not very intense, which makes it possible to tell it is not C=O stretching band. Other stretching bands we can look at in such case are the sp^2 hybridized C—H stretches, which will appear just above 3000 cm^{-1}. 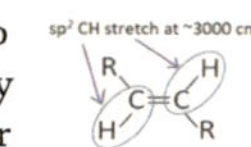
	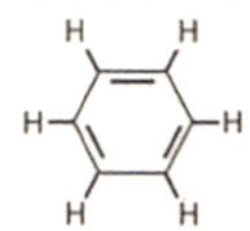	Aromatic C=C bonds stretching	~1610, ~1500 & ~1470	Two to three peaks in the region between 1450 and 1620 cm^{-1} indicate the C=C stretching in aromatic ring, the one at the lowest wavelength (~1470 cm^{-1}) may occur at very small intensity.
C=N		Stretch	1650–1550	
C=O		C=O stretching	1750–1850	Very strong and generally appears between 1650 and 1850 cm^{-1}. If the carbonyl is in conjugation with a double or triple bond, the stretching frequency of the carbonyl will be lowered by 20 to 40 wavenumbers due to the resonance, giving partial single bond character. Example: the carbonyl stretching of cyclohexanone is 1710 cm^{-1}, while the stretching of carbonyl in cyclohexanone is 1690 cm^{-1}. 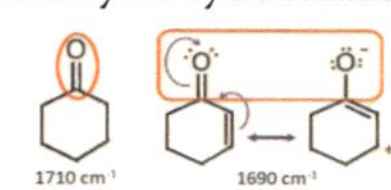
	Ester 		1700–1750	
	Aldehyde 		1720–1740	
	Ketone 		1680–1750	
	Amide 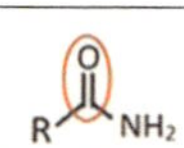		1650–1700	

Bond	Detailed type of bond		Detailed vibration type	Wavenumber (cm^{-1})	Intensity, appearance, and other peaks/pattern information for the identification of the functional groups
C—O	In alcohol	R—OH	Csp^3–O stretching	~1050	The C—O bonds in alcohol and ether are purely single, whereas CO bond in carboxylic acid has a partial double bond character. On the other hand, ester has one C—O that is purely single and another one that exhibits partial double bond character giving rise to two peaks at ~1050 and ~ 1250 cm^{-1}.
	In ether	R—O—R			
	In ROOH	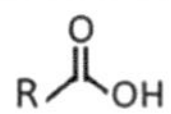	Csp^2–O stretching	~1250	
	In ester	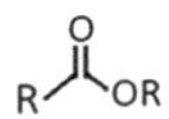	Csp^3–O stretching & Csp^2–O stretching	~1050 and ~ 1250	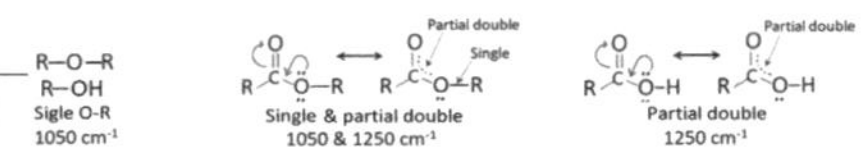

Table 1.
Wavenumber and nature of peaks for different functional groups.

band at about 890 cm^{-1} shows the existence of β-configuration, while the absorption bands at around 840 cm^{-1} or 920 cm^{-1} reveal the presence of α-configuration [4, 5]. On the other hand, IR spectroscopy was used to determine the presence of n-alkane and long-chain alcohol in plant species. Ferreira et al. found that in the presence of intense peaks in the range 1190–1485 cm^{-1}, many elongations such as (C—C), (C—O) coupled to (CH) and (OH) in-plane deformations co-occur, which are associated with n-alkanes and alcohols in the studied samples [6].

AFM-IR (photothermal-induced resonance) was developed and used to characterize four industrially important polymers: a polypropylene-based reactor thermoplastic polyolefin (rTPO), linear low-density polyethylene (LLDPE) for molding, and two recycled post-consumer polypropylene/polyethylene blends. Using nanoscale spatial resolution IR imaging, we could picture each polymer's key components, including the mineral fillers (talc and calcium carbonate) in each mixture and the morphology acquired using AFM. The obtained results show the possibility of individually specifying the polymer and $CaCO_3$ with the peak at 1427 cm^{-1} and the peak at 1459 cm^{-1}, which is related to the CH_2 group vibration and 1376 cm^{-1} absorption band, which corresponds to the CH_3 symmetric vibration band [7].

2.4 Analysis of the inorganic compounds

Simple inorganic compounds, such as NaCl, do not generate vibrations in the MIR range, despite their lattice vibrations in the FIR area; because of this, the IR windows are manufactured of simple inorganic compounds such as NaCl, KBr, and ZnSe. Several variables influence the inorganic compound IR spectra. The compound's crystal structure, like the crystalline lattice bands that appear in the FIR range and changes in the crystalline structure, could be detected from the IR spectra. The IR technique is a nondestructive sampling approach and is preferred for such samples. When analyzing spectra, the degree of hydration of an inorganic substance is also considered. Due to O—H stretching and bending absorption bands, water molecules in a crystalline compound's lattice framework form characteristic sharp bands in the 3800–3200 and 1700–1600 cm^{-1} regions [8]. IR spectroscopy may directly analyze nanoparticles and the functional groups on their surfaces. Furthermore, various ligands attached to nanoparticles may be quickly, precisely, and nondestructively identified using their vibrational fingerprints. In addition to directly analyzing such materials, specific nanostructures alter the local optical field, amplifying IR signals and enabling better-IR nanoparticle imaging processes [9]. According to various investigations, IR spectra indicate vibrational fingerprints of synthesized nanoparticles. On the surface of persimmon peel biochar, the IR spectra of the Si—O vibration can be detected, and the absorption band emerges at 1029 cm^{-1} [10]. Also, silicate (Si—O—Si) deformation vibrations appear on the surface of olive tree leaves biochar in the range 1120–950 cm^{-1} [11], and tetrahedral complexes (Mn^{2+}—O^{2-}) can occur at 700–500 cm^{-1} [12]. Fe—O bond vibration can be shown in the IR spectrum of magnetite nanoparticles, appearing in the range 564–570 cm^{-1} [13, 14]. Finally, bending vibration modes of Ti—OH and Ti—O stretching modes could appear at 1630 and 1383 cm^{-1}, respectively [15].

3. Applications of IR spectroscopy

FT-IR spectroscopy has found many appreciations in the field of medicine, especially as a diagnostic tool, as well as environmental research including protein and nucleic acid analysis [16–18], environmental analysis, forensics, drug syntheses, and pharmaceutical fields. IR spectroscopy has been extensively used for deep analysis of organic and inorganic compounds based on observations of specific absorption bands. The typical applications of IR spectroscopy are compound-type detection, substitution groups, anomeric carbon structures, and crystal allomorphs. However, because some compounds' structures are so similar, the qualitative characterization of these compounds using IR spectroscopy is complex. Furthermore, the overlap of bands in spectra may obscure certain important information, adding to the difficulty of this study. As a result, qualitatively analyzing different compounds using only one characteristic IR peak may be purely empirical and inaccurate. The applications of IR spectroscopy in different science branches are as follows:

3.1 Application to environmental and biological samples

Food products may be assessed qualitatively and quantitatively using MIR and NIR spectra. As significant components, foods are complex mixtures of water, proteins, fats, and carbohydrates. Fats and oils have long been studied using IR spectroscopy. Some vegetable oils, such as canola (CaO), maize (CO), soybean (SO), and walnut (WO), have a similar color to cod liver oil (CLO), making their detection with the human eye very difficult [19]. As a result, FT-IR spectroscopy, in combination with chemometrics, was used to detect vegetable oils. The IR spectra of all these samples show the modest changes in band intensities and the specific frequencies at which maximal absorbance is obtained in each oil due to variances in fatty acid content, chain length, degree, and location of double bonds in triglyceride [19]. These spectrum variations may be found in frequency ranges of 3007 (cis-vinylic stretching vibrations), 2922, and 2852 cm^{-1} due to asymmetrical and symmetrical —CH_2— stretching vibrations and 1237, 1117, and 1098 cm^{-1} related to C—O stretching vibrations [20]. As a result, these frequency regions were used to differentiate between CLO and other vegetable oils. Due to the consumption of hazardous chemicals, fabricated eggs can cause various human health concerns. As a result, developing technology for identifying fabricated eggs is a top concern for food safety. FT-IR spectroscopy is commonly employed because of its ability to analyze such objects using their optical features. The albumen results show that the wavenumber region between 1650 and 1550 cm^{-1} provides excellent separation between synthetic and actual albumen. Even though absorbance values of natural and created albumen exhibit a peak around 3300 cm^{-1}, those peaks linked to water content were eliminated since the water contents in the synthetic egg are easily modified by the recipe [21].

Evaluating the overall soil quality necessitates estimation of the soil's physical, chemical, and biological characteristics. MIR spectroscopy is a rapid, low-cost, environmentally friendly, nonhazardous, and reproducible approach and can be used in different applications as shown in **Figure 2**. The C and N content of the soil was thoroughly investigated using MIR spectroscopy. MIR spectroscopy may also illustrate and assess clay minerals, soil water (volumetric and gravimetric water content), and soil microorganisms [22]. NIR spectroscopy has emerged as a popular method for measuring heavy metals and other pollutants in soils, sediments, and water [23]. Multi-molecular IR (MM-IR) spectroscopy was used to identify and quantify the amount of fluorescent whitening agent in the wheat flour to check the quality of the flour [24]. MM-IR integrated with two-trace 2D correlation spectroscopy was used to analyze deoxynivalenol (DON). DON is considered a potent pollutant in wheat flour, even in low quantities, and causes a range of adverse health impacts on people and cattle [25].

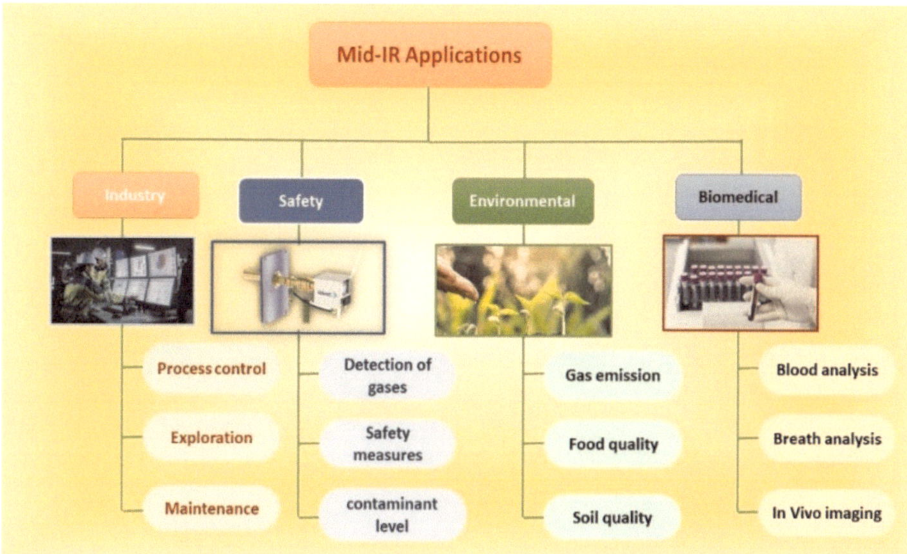

Figure 2.
Applications of MIR spectroscopy.

3.2 Applications to biomedical and bioanalytical investigations

IR spectroscopy has become one of the most powerful tools in modern bioscience. Because of its high molecular specificity, applicability to a wide range of samples, rapid measurement, and noninvasiveness, IR can provide important qualitative and quantitative information for various types of biological material. IR became a well-established bioanalytical approach with several applications [26, 27]. It can be used as an efficient tool for predicting protein secondary structure. One of the most significant obstacles to estimating protein structure using FT-IR spectroscopy is interference from H_2O and the protein IR amide I and II bands. Earlier research used D_2O solution for protein and peptide spectroscopy since there are no D_2O absorption spectra in the area where the amide I and II bands are visible. Later, using a drying apparatus and a tiny sample cell (<10 μm) enabled the quantification of H_2O. Correlations were established between various protein secondary structures in H_2O solution and their FT-IR spectroscopy bands [28]. IR spectroscopy was also used to characterize various systems, including structure elucidation and identification, crystalline and amorphous forms, and quantitative and remote sensing applications [29]. Hence, it can be used to determine roxithromycin in its formulations using conventional KBr to acquire the FT-IR spectra of standards and samples. Based on Beer's law, a calibration curve with (R^2 = 0.9992) was created using the FT-IR carbonyl region (C=O) range 1765–1705 cm^{-1} [30]. In addition, the early detection of breast cancer was done using FT-IR spectroscopy. The ranges 900–1425 cm^{-1} and 1475–1710 cm^{-1} displayed better classification performance. The lipids, proteins, sugars, and nucleic acids were represented by the region 900–1425 cm^{-1}, while the proteins were shown by the region 1475–1710 cm^{-1}. The biological compounds in the other bands also added some unique potential to the categorization, resulting in the best classification accuracy in the entire band [31]. In addition, research showed that integrating imaging technique IR spectroscopy can be promising apparatus for the early diagnosis of diseases such as cancer [32–35] and for monitoring brain stroke damages after treatment [36–38].

Author details

Marwa El-Azazy*, Ahmed S. El-Shafie and Khalid Al-Saad
Department of Chemistry and Earth Sciences, College of Arts and Sciences, Qatar University, Doha, Qatar

*Address all correspondence to: marwasaid@qu.edu.qa

References

[1] Hong T, Yin J-Y, Nie S-P, Xie M-Y. Applications of infrared spectroscopy in polysaccharide structural analysis: Progress, challenge and perspective. Food Chemistry: X. 2021;**12**:100168. DOI: 10.1016/j.fochx.2021.100168

[2] Canteri MHG, Renard CMGC, Le Bourvellec C, Bureau S. ATR-FTIR spectroscopy to determine cell wall composition: Application on a large diversity of fruits and vegetables. Carbohydrate Polymers. 2019;**212**: 186-196. DOI: 10.1016/j.carbpol.2019.02.021

[3] Caputo HE, Straub JE, Grinstaff MW. Design, synthesis, and biomedical applications of synthetic sulphated polysaccharides. Chemical Society Reviews. 2019;**48**(8):2338-2365. DOI: 10.1039/C7CS00593H

[4] Molaei H, Jahanbin K. Structural features of a new water-soluble polysaccharide from the gum exudates of Amygdalus scoparia Spach (Zedo gum). Carbohydrate Polymers. 2018;**182**: 98-105. DOI: 10.1016/j.carbpol.2017.10.099

[5] Zou Y, Du F, Hu Q, Wang H. The structural characterization of a polysaccharide exhibiting antitumor effect from Pholiota adiposa mycelia. Scientific Reports. 2019;**9**(1):1724. DOI: 10.1038/s41598-018-38251-6

[6] Ferreira L, Machado N, Gouvinhas I, Santos S, Celaya R, Rodrigues M, et al. Application of Fourier transform infrared spectroscopy (FTIR) techniques in the mid-IR (MIR) and near-IR (NIR) spectroscopy to determine n-alkane and long-chain alcohol contents in plant species and faecal samples. Spectrochimica Acta Part A: Molecular and Biomolecular Spectroscopy. 2022; **280**:121544. DOI: 10.1016/j.saa.2022.121544

[7] dos Santos ACVD, Lendl B, Ramer G. Systematic analysis and nanoscale chemical imaging of polymers using photothermal-induced resonance (AFM-IR) infrared spectroscopy. Polymer Testing. 2022;**106**:107443. DOI: 10.1016/j.polymertesting.2021.107443

[8] Nakamoto K. Infrared and Raman spectra of inorganic and coordination compounds, part B, applications in coordination. In: Organometallic and Bioinorganic Chemistry. New York: Wiley; 1997

[9] López-Lorente ÁI, Mizaikoff B. Recent advances on the characterization of nanoparticles using infrared spectroscopy. TrAC Trends in Analytical Chemistry. 2016;**84**(Part A):97-106. DOI: 10.1016/j.trac.2016.01.012

[10] Gao L, Wang L, Cao Y, Li S. Persimmon peel-based ion-imprinted adsorbent with enhanced adsorption performance of gallium ions. Minerals Engineering. 2022;**176**:107354. DOI: 10.1016/j.mineng.2021.107354

[11] Mahyoob W, Alakayleh Z, Abu Hajar HA, Al-Mawla L, Altwaiq AM, Al-Remawi M, et al. A novel co-processed olive tree leaves biomass for lead adsorption from contaminated water. Journal of Contaminant Hydrology. 2022;**248**:104025. DOI: 10.1016/j.jconhyd.2022.104025

[12] Yin K, Wang J, Zhai S, Xu X, Li T, Sun S, et al. Adsorption mechanisms for cadmium from aqueous solutions by oxidant-modified biochar derived from Platanus orientalis Linn leaves. Journal of Hazardous Materials. 2022;**428**:128261. DOI: 10.1016/j.jhazmat.2022.128261

[13] El-Azazy M, El-Shafie AS, Al-Meer S, Al-Saad KA. Eco-structured adsorptive removal of Tigecycline from wastewater: Date pits' biochar versus the magnetic biochar. Nanomaterials. 2021; **11**:30. DOI: 10.3390/nano11010030

[14] El-Azazy M, Nabil I, Hassan SS, El-Shafie AS. Adsorption characteristics of pristine and magnetic olive stones biochar with respect to Clofazimine. Nanomaterials. 2021;**11**:963. DOI: 10.3390/nano11040963

[15] El-Azazy M, El-Shafie AS, Morsy H. Biochar of spent coffee grounds as per Se and impregnated with TiO_2: Promising waste-derived adsorbents for Balofloxacin. Molecules. 2021;**26**:2295. DOI: 10.3390/molecules26082295

[16] Militello V, Arluison V. Application of FTIR spectroscopy to analyze RNA structure. Methods Mol Biol, Frédéric Geinguenaud. 2020;**2113**: 119-133. DOI: 10.1007/978-1-0716-0278-2_10

[17] Han Y, Ling S, Qi Z, Shao Z, Chen X. Application of far-infrared spectroscopy to the structural identification of protein materials. Physical Chemistry Chemical Physics. 2018;**20**(17):11643-11648. DOI: 10.1039/c8cp00802g

[18] El-Azazy M. Introductory chapter: Infrared spectroscopy - a synopsis of the fundamentals and applications. In: El-Azazy M, editor. Infrared Spectroscopy - Principles, Advances, and Applications. London: IntechOpen; 2018. DOI: 10.5772/intechopen.82210

[19] Rohman A, Che Man YB. Application of Fourier transform infrared (FT-IR) spectroscopy combined with chemometrics for authentication of cod-liver oil. Vibrational Spectroscopy. 2011;**55**(2):141-145. DOI: 10.1016/j.vibspec.2010.10.001

[20] Guillén MD, Cabo N. Infrared spectroscopy in the study of edible oils and fats. Journal of the Science of Food and Agriculture. 1997;**75**:1-11

[21] Joshi R, Baek I, Joshi R, Kim MS, Cho B-W. Detection of fabricated eggs using Fourier transform infrared (FT-IR) spectroscopy coupled with multivariate classification techniques. Infrared Physics & Technology. 2022;**123**:104163. DOI: 10.1016/j.infrared.2022.104163

[22] Nath D, Laik R, Meena VS, Kumari V, Singh SM, Pramanick B, et al. Strategies to admittance soil quality using mid-infrared (mid-IR) spectroscopy an alternate tool for conventional lab analysis: A global perspective. Environmental Challenges. 2022;**7**:100469. DOI: 10.1016/j.envc.2022.100469

[23] Cozzolino D. Near infrared spectroscopy as a tool to monitor contaminants in soil, sediments and water—State of the art, advantages and pitfalls. Trends in Environmental Analytical Chemistry. 2016;**9**:1-7. DOI: 10.1016/j.teac.2015.10.001

[24] Pan Q, Xie J, Lin L, Hong M-S, Wang X-C, Sun S-Q, et al. Direct identification and quantitation of fluorescent whitening agent in wheat flour based on multi-molecular infrared (MM-IR) spectroscopy and stereomicroscopy. Spectrochimica Acta Part A: Molecular and Biomolecular Spectroscopy. 2021;**250**:119353. DOI: 10.1016/j.saa.2020.119353

[25] Li F-L, Xie J, Wang S, Wang Y, Xu C-H. Direct qualitative and quantitative determination methodology for massive screening of DON in wheat flour based on multi-molecular infrared (MM-IR) spectroscopy with 2T-2DCOS. Talanta. 2021;**234**:122653. DOI: 10.1016/j.talanta.2021.122653

[26] Leal LB, Nogueira MS, Canevari RA, Carvalho LFCS. Vibration spectroscopy

and body biofluids: Literature review for clinical applications. Photodiagnosis and Photodynamic Therapy. 2018;**24**:237-244. DOI: 10.1016/j.pdpdt.2018.09.008

[27] Beć KB, Grabska J, Huck CW. Biomolecular and bioanalytical applications of infrared spectroscopy – A review. Analytica Chimica Acta. 2020; **1133**:150-177. DOI: 10.1016/j.aca.2020.04.015

[28] Yang S, Zhang Q, Yang H, Shi H, Dong A, Wang L, et al. Progress in infrared spectroscopy as an efficient tool for predicting protein secondary structure. International Journal of Biological Macromolecules. 2022;**206**:175-187. DOI: 10.1016/j.ijbiomac.2022.02.104

[29] Eerdenbrugh BV, Taylor LS. Application of mid-IR spectroscopy for the characterization of pharmaceutical systems. International Journal of Pharmaceutics. 2011;**417**(1–2):3-16. DOI: 10.1016/j.ijpharm.2010.12.011

[30] Sherazi STH, Ali M, Mahesar SA. Application of Fourier-transform infrared (FT-IR) transmission spectroscopy for the estimation of roxithromycin in pharmaceutical formulations. Vibrational Spectroscopy. 2011;**55**(1):115-118. DOI: 10.1016/j.vibspec.2010.09.010

[31] Du Y, Xie F, Yin L, Yang Y, Yang H, Wu G, et al. Breast cancer early detection by using Fourier-transform infrared spectroscopy combined with different classification algorithms. Spectrochimica Acta Part A: Molecular and Biomolecular Spectroscopy. 2022;**283**:121715. DOI: 10.1016/j.saa.2022.121715

[32] Ali MH, Rakib F, Al-Saad K, Alsaady R, Lyng FM, Goormaghtigh E. A simple model for cell type recognition using 2D-correlation analysis of FTIR images from breast cancer tissue. Journal of Molecular Structure. 2018;**1163**(5):472-479. DOI: 10.1016/j.molstruc.2018.03.044

[33] Lyng FM, Traynor D, Nguyen TNQ, Meade AD, Rakib F, Al-Saady R, et al. Discrimination of breast cancer from benign tumours using Raman spectroscopy. PLoS One. 2019;**14**(2): e0212376. DOI: 10.1371/journal.pone.0212376

[34] Ali MHM, Rakib F, Al-Saad K, Al-Saady R, Goormaghtigh E. An innovative platform merging elemental analysis and Ftir imaging for breast tissue analysis. Scientific Reports. 2019;**9**(1):9854. DOI: 10.1038/s41598-019-46056-4

[35] Ali MHM, Toor SM, Rakib F, Mall R, Ullah E, Mroue K, et al. Investigation of the effect of PD-L1 blockade on triple negative breast cancer cells using Fourier transform infrared spectroscopy. Vaccine. 2019;**7**(3):109. DOI: 10.3390/vaccines7030109

[36] Rakib F, Ali CM, Yousuf M, Afifi M, Bhatt PR, Ullah E, et al. Investigation of biochemical alterations in ischemic stroke using Fourier transform infrared imaging spectroscopy—A preliminary study. Brain sciences. 2019;**9**(11):293. DOI: 10.3390/brainsci9110293

[37] Ali MH, Al-Saad K, Popelka A, van Tilborg G, Goormaghtigh E. Application of Fourier transform infrared (FTIR) spectroscopy and atomic force microscopy in stroke-affected brain tissue. Swift Journal of Medicine and Medical Sciences. 2016;**2**:011-024

[38] Ali M, Rakib F, Abdelalim EM, Mall R, Ullah E, Mesaeli N, et al. Fourier-transform infrared imaging spectroscopy and laser ablation-ICPMS new vistas for biochemical analyses of ischemic stroke in rat brain. Frontiers in Neuroscience. 2018;**12**:647. DOI: 10.3389/fnins.2018.00647

Chapter 2

Attenuated Total Reflectance Mode for Transport through Membranes

Daniel T. Hallinan Jr

Abstract

This chapter is an introductory tutorial to attenuated total reflectance (ATR) mode of Fourier-transform infrared spectroscopy and how it can be used to measure transport through polymer membranes. In addition to covering the experimental set-up and time-resolved data processing, it will present the fundamental equations for analyzing the data in order to obtain diffusion coefficients. The chapter will present several example systems in which FTIR-ATR has been used to determine transport, including water diffusion through polyelectrolytes for fuel cells and block copolymers for water purification as well as ion transport through polymer electrolytes for lithium batteries. Perspectives on future applications in which the technique could provide fundamental understanding will also be covered.

Keywords: FTIR-ATR, ATR, membrane transport, polymer, diffusion, time-resolved, water purification, desalination, separations, batteries

1. Introduction

Fourier transform infrared spectroscopy (FTIR) revolutionized chemical analysis with light-based spectroscopy. The ability to pass all wavelengths of mid-infrared light through a sample simultaneously rather than one wavelength at a time increased sampling rate by orders of magnitude. The IR source beam is split into a reference beam that reflects off a moving a mirror, thereby changing the path length in time, and a beam that passes through the sample. FTIR takes advantage of constructive and destructive interference between the reference and sample beams that is referred to as an interferogram. Due to the interference changing in time, the interferogram can be converted into intensity versus wavenumber via a Fourier transform. Wavenumber is inversely proportional to wavelength and the mid-IR region is from 4000 to 400 cm^{-1}. Intensity is converted to transmission by dividing the sample measurement by a background, which is a measurement without a sample present. This results in an infrared spectrum with a baseline at or near 100% in wavenumber regions where no light is absorbed by the sample, i.e. where the frequency of light does not correspond to the frequency of vibration of any of the molecular functional groups of the sample. In wavenumber regions where light is absorbed a "peak" is observed. The peak is typically a Gaussian decrease in transmitted intensity centered around the frequency of a particular vibration of a functional group containing covalent bond(s). The combination of functional groups in a sample results in a spectrum that can be used to identify the chemical make-up of unknown samples or detect the presence of particular species. Thus, FTIR provides a rapid and reliable means for sample identification and/or detection. Greater detail regarding the fundamentals of FTIR, prediction of functional group absorbance, and tables of wavenumber values can be found elsewhere [1].

One of the limitations of FTIR is that if samples are too thick, an insufficient amount of light passes through the sample for detection. This is due both to absorbance and scattering. This can be overcome by preparing thin samples, on the order of microns, or, if the sample is a powder, it can be mixed with a salt, such as KBr, and pressed into a transparent pellet. Inorganic salts contain only ionic bonds that do not absorb infrared light. So, if the sample is dilute in the pellet, then sufficient transmission is achieved. An alternative to sample processing is to use attenuated total reflectance (ATR) to ensure that sufficient light intensity is transmitted. As shown in **Figure 1**, the infrared beam is directed into the crystal at an angle less than the critical angle, typically 0° with respect to the normal of the face through which it enters. The beam then totally internally reflects at the top surface of the crystal because the incident angle is greater than the critical angle. The critical angle is a function of ratio of the index of refraction of the crystal and the medium above the reflecting surface. Crystals of high refractive index are commonly used, such that the critical angle is small. For ATR set-ups without adjustable angles of incidence, the incident angle on the top face is usually 45° as shown in **Figure 1**. After reflecting at the top surface of the crystal, the infrared beam exits the crystal and is directed to the detector. The crystal is housed on an optical set-up that directs the source beam to the crystal and the exiting beam to the detector with a combination of mirrors and sometimes focusing lenses. The mirrors and lenses are adjustable so that the amount of infrared energy transmitted to the detector (without a sample present) can be maximized.

Common materials of construction for ATR crystals include diamond, ZnSe, KRS-5, silicon, and germanium. Despite the cost, diamond is an excellent ATR material due to its robustness. It is stable to high temperature, in contact with corrosive substances (including both high and low pH), and in contact with abrasive samples. Other than cost, which limits diamond ATR crystals to single reflection, the only other drawback is its inherent absorbance of IR between approximately 2700 and 1800 cm^{-1}. Although noise is greater in this region of the spectrum, this is actually only a minor drawback because the diamond absorbance is part of the background and thus subtracted from the sample spectrum. Moreover, there are few important sample peaks found in this wavenumber range. If robustness is not important, then the best

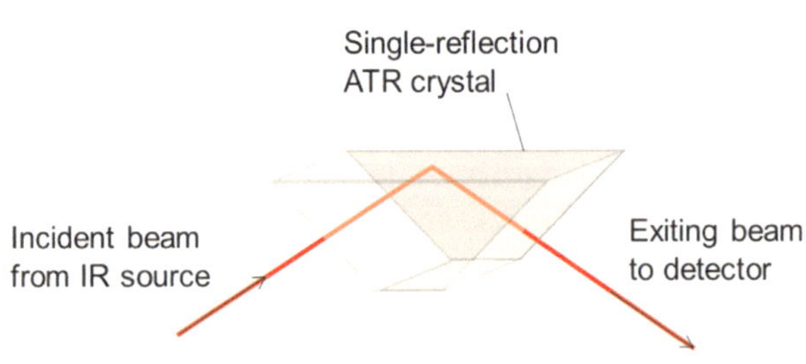

Figure 1.
Schematic of an ATR crystal with an infrared beam entering from the source side, totally internally reflecting, and exiting to the detector. In this schematic the angle of incidence is 45°, a common fixed incidence angle, and there is a single reflection. Not shown are the FTIR spectrometer, the mounting that houses the crystal, the sample that is placed on top of the crystal, nor the anvil that presses the sample to the crystal ensuring intimate contact which is essential for high-quality reproducible data collection.

choice for maximizing throughput of infrared energy is ZnSe. This material is available as single-reflection crystals as well as multiple reflection crystals that increase the sensitivity of detection. Commercial KRS-5 is rarely used today due to toxicity. Silicon and germanium have intermediate properties (including cost and energy transmission) for select applications. For example, due to lower infrared energy transmission, germanium can be used to solve absorption saturation problems that sometimes occur in multi-reflection ZnSe experiments.

ATR mode obviates the need for sample preparation because sampling occurs via an evanescent wave that exists outside the crystal and therefore within the sample. The evanescent wave is a non-propagating (i.e. standing) wave that forms due to interaction between the incident and reflected wave. Although it is non-propagating, the evanescent wave can interact with the sample such that infrared energy can be absorbed by the sample and an infrared spectrum generated. The intensity of the evanescent wave is exponentially decaying such that it is most intense at the crystal surface. The depth at which the intensity decays to e^{-1} is defined as the depth of penetration, d_p. It is determined by the physics of the total internal reflection. For reflection at an interface between two dielectric (i.e. non-metallic) materials it is a function of the angle of incidence, θ, refractive index ratio of crystal, n_{ATR}, and sample, n_{sample}, and wavelength of light, λ [2].

$$d_p = \frac{\lambda}{2\pi n_{ATR}\sqrt{\sin^2(\theta) - \left(\frac{n_{sample}}{n_{ATR}}\right)^2}} \tag{1}$$

For internal reflection to occur $n_{sample} < n_{ATR}$. As shown in **Table 1**, the refractive indices of air, water, and polymers are all less than that of materials commonly used to construct ATR crystals. By way of example, the critical angle for total reflection within diamond with water on top is $\theta_c = \sin^{-1}\left(\frac{n_{sample}}{n_{ATR}}\right) = 33°$. This is measured from the normal to the interface, such that $\theta > \theta_c$ for total internal reflection. Keeping with this example, at the highest wavenumber (4000 reciprocal centimeters) $d_p = 0.4$ µm. Due to practical limitations of most mid-infrared optics, the lowest wavenumber at which reliable data can be obtained is 650 cm^{-1}. At this wavenumber, $d_p = 2.3$ µm. This demonstrates that the only significant drawback to ATR mode is that the sampling

Material	Refractive Index	Reference
Air	1.0	[3]
Water	1.3	[3]
Polymer	~1.5	[4]
Diamond	2.38	[5, 6]
ZnSe	2.45	[3]
Silicon	3.4	[7]
Germanium	4.0	[7]

Table 1.
Approximate refractive indices of selected samples and ATR materials at room temperature and at wavenumbers between 4000 and 650 cm^{-1} (wavelengths between 2.5 and 15 µm).

depth is a function of wavenumber, which causes a slight skewing of relative intensity of different peaks when compared to transmission mode. Since most libraries of infrared spectra have been constructed in transmission mode, it is best practice to use transmission mode for identification of unknown compounds. However, the aforementioned skewing of relative peak intensities is only slight, such that library searches from ATR spectra are usually successful. Moreover, instrument software often has automated algorithms to correct for this effect, which can also be rigorously accounted for manually if absolute peak intensities are required.

On the other hand, there are several benefits to ATR mode that outweigh the wavenumber dependence of d_p in most cases. Regardless of the total sample thickness, infrared is absorbed over only about $5d_p \sim 10$ µm at most. Thus, sufficient energy transmission to the detector occurs without any special sample preparation for all but rare instances, such as carbon black that strongly absorbs and scatters infrared light. In addition to ease of sample preparation, temperature is readily controlled in the ATR set-up as is the environment of the sample. Thus, time-resolved FTIR-ATR measurements can be used for reaction kinetics and membrane transport studies, the focus of this chapter.

2. Fundamentals of ATR

As with transmission mode FTIR, a background must be collected when using ATR mode. It is essential that the ATR accessory is in place when collecting the background so that absorption by the ATR crystal is included in the background spectrum. For this reason, the appearance of the background will depend on type of ATR crystal being used. Best practice is for the top surface of the crystal to be empty and dry when collecting the background, although it is possible to have the sample in place and use the initial conditions of the sample as the background in order to generate difference spectra if the main purpose of the experiment is to examine changes in time. This latter approach is not recommended because of the detrimental impact it has on signal-to-noise ratio [3].

After the background has been collected, the sample is placed on top of the crystal. Due to the exponential decay of the evanescent wave intensity, intimate contact between sample and crystal is crucial. This is trivial for liquid samples, but an anvil that generates reproducible force should be used to press solid samples to the crystal surface. An alternative approach that has been used for polymeric samples is to cast a polymer film from solution directly onto the ATR crystal. This approach was used in the pioneering work in which FTIR-ATR spectroscopy was first demonstrated as a powerful technique for quantifying transport in polymer films and is covered in an excellent review [8]. Layer-by-layer deposition has also been used to assemble polymer films on ATR crystals [9]. Care should be taken with polyelectrolyte solutions that can exhibit highly acidic or basic conditions that will damage crystals such as ZnSe. In those cases, diamond can be used, the sample can be physisorbed to the crystal [10], or a pressure-contact method can be utilized [11].

An important consideration in ATR mode is the thickness of the sample. As shown in **Figure 2(a)**, the evanescent wave will extend beyond a thin sample of less than a few d_p. For transport experiments in which the conditions at the top surface of the sample are being changed, this is a problem because the functional groups present throughout the sample and outside the sample are all being measured, weighted by the exponential decay of the evanescent wave. This can be expressed quantitatively using

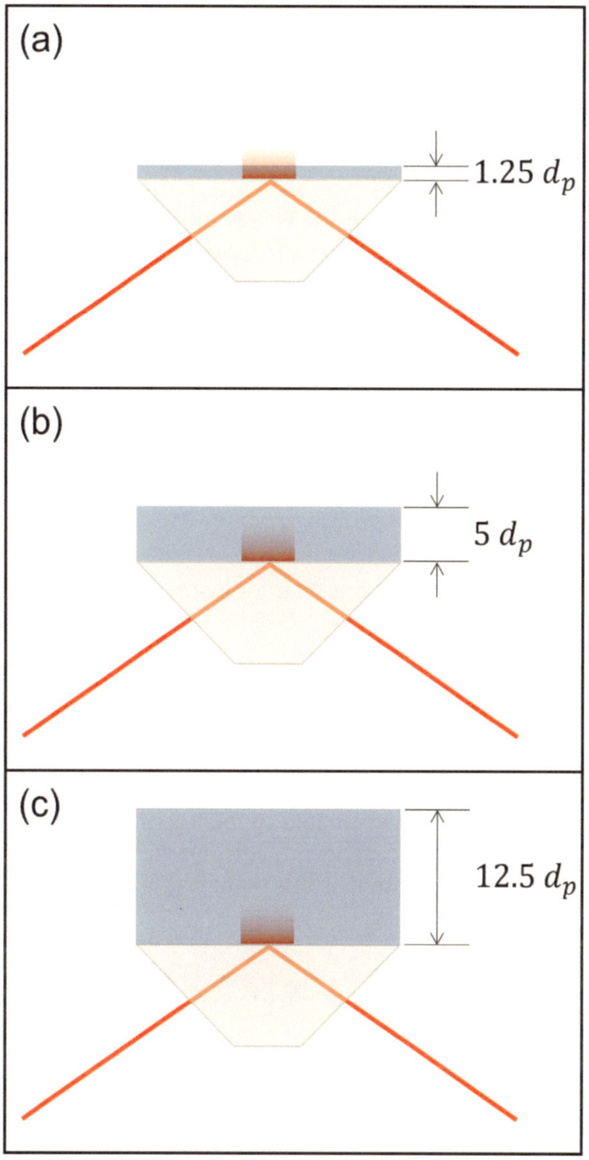

Figure 2.
Schematics of evanescent wave and relative thickness of sample for (a) a sample that is less than the ideal thickness for ATR, (b) a sample in which the evanescent wave extends throughout the sample, and (c) a sample much thicker than the depth of penetration in which detection can be approximated as occurring at the crystal surface.

the Beer-Lambert Law, which states that for weak to moderate absorption, absorbance (A) is proportional to molar concentration (C) [12],

$$A = -\log\left(\frac{I}{I_0}\right) = \varepsilon CL \tag{2}$$

Note that I is the measured intensity, I_0 is the intensity of the background, ε is the molar extinction coefficient, and L is the path length of the IR beam through the sample. For ATR mode, the decay of the evanescent wave energy should be incorporated, $I = E^2 = I_0 e^{-\frac{2y}{d_p}}$.

This has been achieved by using the differential form of the Beer-Lambert Law in terms of intensity [12], but it assumes weak absorbance, which introduces 10% error at 80% transmission and greater error with decreasing transmission (increasing absorbance). Moreover, the original reference used natural logarithm to define absorbance, rather than the typical log base 10. This approximation was necessary to yield an analytical solution to transient diffusion [8], but alternate approaches are possible and discussed below. In any case, an exact expression is derived here.

$$A = -\log\left(\frac{I}{I_0}\right) \tag{3}$$

Relating dA to dI and inserting into Beer-Lambert Law yields.

$$dA = -\frac{dI}{I\, ln\, 10} = \varepsilon C\, dy \tag{4}$$

Rearranging and inserting the expression for evanescent wave decay of intensity yields.

$$dI = -\,ln\, 10\ \varepsilon CI\, dy = -\,ln\, 10\ \varepsilon C I_0 e^{-\frac{2y}{d_p}}\, dy \tag{5}$$

The left-hand side of the equation can be integrated from the incident intensity, I_0, to I. Due to the exponential decay, the integral of the right-hand side can be evaluated from 0 to infinity, but is typically evaluated to the top of the sample, L, because beyond this the molar extinction coefficient, ε, is not well defined. In either case, the integral cannot be explicitly evaluated at this point if C is a function of y, which will be the case in transport measurements.

$$\frac{I - I_0}{I_0} = -\,ln\, 10 \int \varepsilon C(y) e^{-\frac{2y}{d_p}}\, dy \tag{6}$$

In terms of absorbance, this is

$$A = -\log\left(1 - ln\, 10 \int \varepsilon C(y) e^{-\frac{2y}{d_p}}\, dy\right) \tag{7}$$

For a known concentration profile, i.e. known $C(y)$, the integral can be evaluated and an explicit expression for absorbance found. Alternatively, the integral can be evaluated numerically if necessary.

The case of **Figure 2(a)** where the sample is thinner than a few d_p is not ideal due to the integral in Eq. (7) needing to be evaluated from 0 to L for sample absorption and from L to infinity for absorption by the medium beyond the sample. There will be much less experimental uncertainty for the case of **Figure 2(b)** in which the evanescent wave decays by more 99% of I_0 within the sample. This means that IR absorption beyond the sample is negligible. However, if there is a gradient in concentration of the absorbing species within the sample, such that C is a function of y, then the appropriate expression for $C(y)$ must be incorporated before evaluating the integral of Eq. (7). In other words, a transport model should be incorporated into Eq. (7) or an approximation thereof in order to accurately predict the rate of change of A. The ideal situation, at least for simplicity in transport modeling, is the case of **Figure 2(c)** in which the sample is at least an order of magnitude thicker than d_p. Since 95% of the evanescent wave energy decays within 3 d_p, detection is essentially occurring only at the crystal sample interface for thick samples. This approximation works reasonably well when the sample that is at least 10 d_p, and is basically exact for samples of greater than 100 d_p . In this case, the absorbance is proportional to concentration at $y = 0$, and the integral can be directly evaluated yielding.

$$A = -\log\left[1 - \frac{ln\,(10)\varepsilon C(0)d_p}{2}\right] \tag{8}$$

An effective ATR extinction coefficient can be used to simplify the expression, $b = \frac{ln\,(10)\varepsilon d_p}{2}$.

$$A = -\,log\,[1 - bC] \tag{9}$$

This demonstrates that the choice of base for the logarithm is arbitrary, but it should be specified. In the limit as concentration, and hence absorbance, go to zero (as assumed by Fieldson and Barbari) [12], Eq. (9) further simplifies to $A = bC$. Thus, in the thick film limit depicted in **Figure 2(c)**, absorbance is proportional to concentration in the limit of weak IR absorption.

3. Membrane transport

3.1 Fickian diffusion

Transport in membranes and polymer films is relevant to a wide range of applications that include barriers, electrolytes, and membrane separations. In the area of barriers, polymer films are used in food packaging to improve quality, extend lifetime, and reduce waste. In these applications tailoring transport of water vapor, oxygen, and other gases like ethylene are important as they control the rate of ripening and spoiling. Polymer films are used as building wraps to exclude humidity and control mold. They are used in a wide range of packing applications to protect products, e.g. to maintain sterilization. In the area of electrolytes, polymers are used as solid electrolytes in hydrogen fuel cells, in batteries, and in dialysis and reverse electrodialysis. In the area of separations, porous membranes are used for filtration. Dense polymer membranes are used as reverse osmosis membranes for desalination. Glassy polymers with high free-volume have proven to perform well in separating gases. In essentially all these applications the rate of transport of different species, mostly small molecules and ions, are important.

Transport of water vapor, gases, and ions can be driven by gradients. One of the most common gradients is created by a pressure difference between the opposite sides of the membrane. This pressure difference creates a chemical potential gradient within the membrane, driving transport from the high pressure side to the low pressure side [13]. Although different external gradients can be applied, the universal thermodynamic driving force for transport is that of the electrochemical potential. A gradient of electrochemical potential can be generated by a pressure difference (as already stated), by a concentration gradient, by a voltage gradient, and by a temperature gradient, to name several of the most common driving forces employed in membrane separations. Pressure, temperature, and concentration gradients will universally drive diffusion of all species, but voltage gradients will only drive transport of charged species and their associated solvation shell. This latter mode of transport is commonly referred to as migration for ions and electroosmotic drag for the ions' solvation shell.

Water is by far the most common diffusant in studies with FTIR-ATR. Since 2000, its diffusion has been examined in asphalt [14], in plasticized polyvinylchloride [15], in cellulose acetate [16], in poly(ethylene terephthalate) [17], in polylactide [18], in polystyrene-poly(isobutylene)-polystyrene block copolymers [19], in fuel cell membranes [10, 20, 21], and in ion-selective membranes for corrosion prevention [22–24]. Alcohols are the next most common diffusant that has been investigated with FTIR-ATR, presumably due to the strong OH stretching absorbance present in both water and alcohols. In particular, there are several studies of methanol diffusion in various types of polymers [10, 12, 17, 25]. In other studies, FTIR-ATR has been used as a probe to measure changing composition in the receptor compartment of permeation experiments for multicomponent alcohol transport through membranes [26–28], but this format is beyond the scope of this chapter because the membrane is not in contact with the ATR crystal. Beyond water and alcohols, there is a report of acetonitrile diffusion in cellulose acetate [29]. Finally, the rate of drug release through synthetic skin membranes has been measured with FTIR-ATR [30, 31].

In membrane transport experiments generally and in experiments with FTIR-ATR specifically, it is typically appropriate to assume that transport is 1D because the thickness of the membrane is much less than the lateral dimensions. As shown in **Figure 3**, we will define the thickness coordinate as y. Although not drawn to scale, the control volume shown in **Figure 3** extends from the crystal surface at $y = 0$ to the top of the sample at $y = L$. Regardless of the material of construction, all ATR crystals are impermeable, which means that a no-flux boundary condition exists at $y = 0$. The boundary condition on the top surface will depend on the particular experiment. A common condition is a constant concentration, which can be imposed by placing or flowing a fluid (liquid or vapor) across the top surface of the membrane [8]. If a gradient is imposed in a different way, such as by stacking two membranes of different concentration, then the top boundary condition could also be no flux [32]. Another exciting application of FTIR-ATR is spectroelectrochemistry, in which the

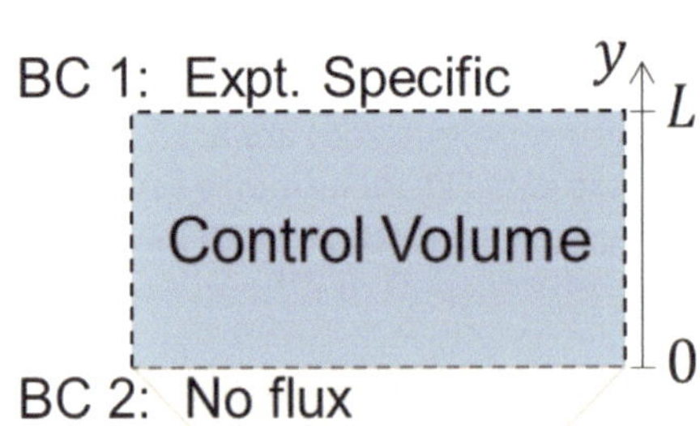

Figure 3.
Schematic of coordinate system, definition of control volume, and designation of boundary conditions (BCs) for a one dimensional transport experiment conducted with an ATR set-up.

driving force for transport is applied potential [33]. If blocking electrodes are used, as in studies of capacitors, then both boundary conditions would be no flux [34]. In battery studies, flux of the active ion could occur at the boundaries if reversible electrodes are used [35]. The flux would match the rate of electrochemical reaction [36]. For transport of neutral combinations of species, the boundary conditions would be no flux in the battery as well.

The governing equation for the control volume depends on the particular experiment being conducted. For the case of Fickian diffusion, driven by a concentration gradient, the transient diffusion equation applies.

$$\frac{\partial C}{\partial t} = D\frac{\partial^2 C}{\partial y^2} \tag{10}$$

Due to the solid-like nature of polymer membranes and films, convection is neglected in this equation. Due to the finite control volume, this partial differential equation is most readily solved with a Fourier Series solution using Finite Fourier Transforms (FFT). The basis functions for the FFT are chosen to satisfy the type of boundary conditions. With appropriate nondimensionalization, the boundary conditions can be made homogeneous for all examples discussed in this chapter.

$$\theta = \frac{C(y,t) - C(0,0)}{C(0,\infty) - C(0,0)}, \tag{11}$$

$$\eta = \frac{y}{L}, \tag{12}$$

$$\tau = \frac{L^2 t}{D} \tag{13}$$

Focusing on the situation in which the thick film approximation is appropriate, the initial and equilibrium concentration at $y = 0$ are used to make concentration dimensionless. The control volume size, L, and diffusion coefficient, D, are used to define dimensionless position and time. The experimental set-up will dictate the initial conditions throughout the control volume, $C(y, 0)$, which must be known in order to find a solution.

As shown in **Figure 4(a)**, the most common initial condition is the case where the control volume has a homogeneous initial concentration and a different concentration is applied at the top boundary. This is frequently accomplished by way of a flowing stream of liquid, vapor, or gas [8]. Shown in **Figure 4(b)**, another possible initial condition is to introduce two separate polymer membranes each with a different concentration within the control volume. In this case, the initial concentration profile contains a step change. This has been used with polymer electrolytes to study salt diffusion, which was necessary to exclude liquid solvents [32]. This case is more easily modeled numerically, for example with finite difference methods, but the step change can be Fourier transformed. Finally, it is possible to introduce a concentration gradient via an applied external gradient, such as an electric field or a temperature gradient. If the relevant transport parameter(s) are constant, then the equilibrium condition with an electric field or temperature gradient applied should be a linear concentration gradient. The field or applied gradient can then be turned off or removed and Fickian diffusion alone will result in the control volume returning to a homogeneous concentration over time. This initially linear concentration profile is depicted schematically in **Figure 4(c)**. The definition of the dimensionless parameters and the Fourier Series

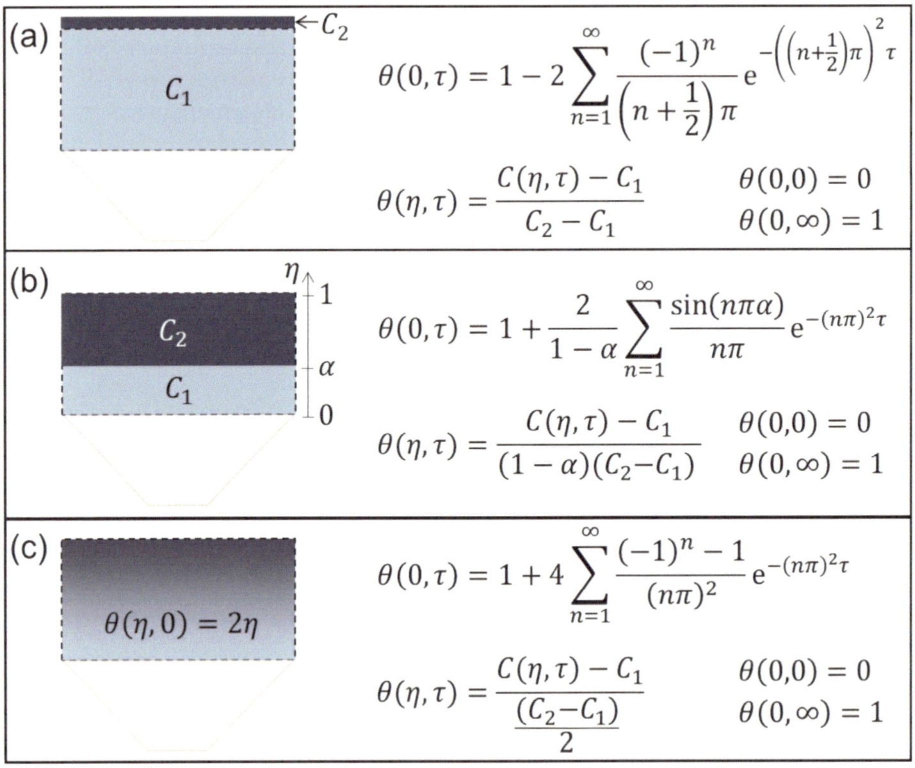

Figure 4.
Possible initial conditions used in FTIR-ATR diffusion studies. In all cases, $\eta = \frac{y}{L}$, $\tau = \frac{L^2 t}{D}$, and the boundary condition at $\eta = 0$ is no flux. (a) There is a homogeneous initial concentration, C_1, throughout the control volume and at $\eta = 1$ an infinite source of concentration C_2. The basis functions for this case are $\phi_n(\eta) = \sqrt{2}\cos\left(\left(n + \frac{1}{2}\right)\pi\eta\right)$. (b) There are two different regions, each of homogeneous concentration. There is a step change of the initial concentration at $\eta = \alpha$. At $\eta = 1$, there is a no-flux boundary condition. The basis functions for this case are $\phi_0 = 1$ for $n = 0$ and $\phi_n = \sqrt{2}\cos(n\pi\eta)$ for $n > 0$. (c) The initial concentration profile is linear from C_1 at $\eta = 0$ to C_2 at $\eta = 1$. The boundary conditions and therefore basis functions for this case are the same as in (b).

solutions at $\eta = 0$ are also reported in **Figure 4** for these three possible diffusion experiments. The solution also as a function of position would include the cosine from the appropriate basis function [37], which goes to one at $\eta = 0$.

The final solutions in **Figure 4** are expressions for concentration as a function of time. Implicit in these expressions are L and D. The former must be known and is assumed to be much larger than the depth of penetration of the evanescent wave, i.e. the thick film approximation. The latter can be used as an adjustable parameter to minimize the error between the model and time-resolved FTIR-ATR data. Thus, the diffusion coefficient can be found from time-resolved FTIR-ATR measurements of concentration gradients dissipating via diffusion. In these experiments, D is a mutual diffusion coefficient for flux of the diffusing species in the polymer matrix under the influence of a concentration gradient. It is not a self or tracer diffusion coefficient, but it can be related to these under certain assumptions and considering the concentration dependence of the activity coefficient [38]. The mutual diffusion coefficient can be concentration dependent, but has been found approximately constant in many membrane systems [39, 40].

3.2 Swelling

In Section 3.1, the thickness of the film was assumed to be constant. For many membrane transport situations this is a good assumption. However, if the membrane absorbs large amounts of penetrant, then the thickness can increase with time (or decrease in the case of desorption). This is most likely to occur in experiments of **Figure 4(a)** in which the average concentration of the control volume is changing with time.

$$\langle C \rangle = \int_0^1 C\, d\eta = \frac{\int_0^L C\, dy}{\int_0^L dy} \tag{14}$$

In experiments of **Figure 4(b)** and **(c)**, $\langle C \rangle$ is constant, but the concentration profile is changing with time. In other words, none of the diffusing species enters or exits the control volume, but its distribution within the control volume does change with time. A thickness change would occur in these cases if there were a significant and concentration-dependent volume change upon mixing between polymer and diffusing species. However, in experiments of **Figure 4(a)** the thickness would change even for the simple case where volume additivity holds, because there is flux into the control volume (or out of it for desorption).

An excellent study of swelling (or dilation) was conducted with FTIR-ATR on several different polymer films with different organic penetrants [41]. Baschetti et al. reported a convenient graphical approach with which to analyze dilation. They plotted the integrated absorbance, which is typically calculated as the area under the peak of interest with a linear baseline subtraction, as shown in the sample time-resolved data of **Figure 5(a)**. This is done at each time point and the value can be plotted versus time, as shown in **Figure 5(b)**. Dilation was assessed by plotting the integrated absorbance of a peak associated with the polymer versus the integrated absorbance of a peak associated with the diffusing species. This plot was found to be linear in rubbery polymers, because dilation proceeds instantaneously with diffusion. On the other hand, a delay in the linear correlation was found in glassy polymers due to

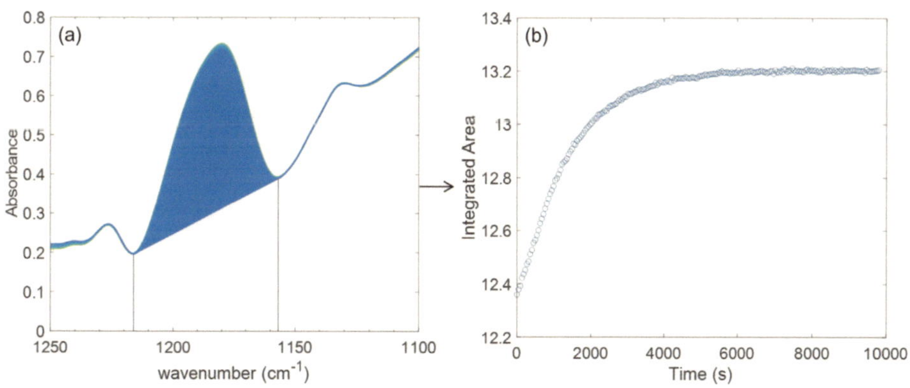

Figure 5.
Example time-resolved data showing the procedure for (a) subtracting a linear baseline and integrating the area under a peak of interest and (b) plotting the integrated area versus time. Only a few representative spectra are shown in (a), but one spectrum was collected for each data point in (b).

polymer relaxation (or plasticization) being slower than diffusion. Several other studies have also considered the effect of polymer relaxation during diffusion, which can result in diffusion appearing non-Fickian [11, 18, 20, 21].

The same integrated absorbance shown in **Figure 5** can be conveniently normalized from zero to one.

$$\frac{A(t) - A(0)}{A(\infty) - A(0)} = \frac{C(t) - C(0)}{C(\infty) - C(0)} \tag{15}$$

In the context of the thick film approximation with weak absorbance, the normalized integrated absorbance is equal to the normalized concentration because the constant of proportionality drops out. There are quantitative approaches to determine if absorbance is weak that are essentially calibration measurements. The simplest approach is to plot absolute absorbance versus known concentrations and look for linearity, but it is often convenient to use ratios of integrated absorbance values in order to correct for dilation or other concentration-dependent phenomena that do not fundamentally violate the weak absorbance assumption [10, 42].

3.3 Non-Fickian diffusion

Non-Fickian diffusion is a misnomer because concentration-gradient-driven diffusion always follows Fick's Law. However, if other phenomena occur during diffusion, then transport can appear to be non-Fickian [43]. A common example is a penetrant diffusing into a glassy polymer and plasticizing it. If the rate of penetrant diffusion within the plasticized polymer is much faster than the rate within the glassy polymer, then Case II diffusion will occur that appears as a sharp front of constant velocity moving through the polymer. The front is the boundary between the glassy polymer and the plasticized polymer. In this case, plasticization of the polymer is rate limiting and it is not diffusion that is being measured at all. General treatment of Case II diffusion has been achieved by treating the plasticization as a swelling [44] or polymer relaxation [45] that is rate limiting.

Another example of apparent non-Fickian diffusion has been observed when the polymer contains groups that interact with the diffusing penetrant. The interaction can be in the form of a reaction, for example a polyelectrolyte reacting with water forming a dissociated hydronium ion and an anionic functional group. This was observed in dry Nafion® that was exposed to water vapor in experiments of **Figure 4(a)** type, where the sulfonate species were dissociated by water [20]. The consumption of mobile water molecules by the reaction caused a delay in the appearance of water molecules at the ATR crystal interface as well as an appearance of a protonated water bending peak in the FTIR spectra as it was formed by the reaction, as shown in **Figure 6**. This study had relevance to hydrogen fuel cells and desalination membranes. Another example of polymer-penetrant interaction causing apparent non-Fickian transport has been observed in semicrystalline poly(ethylene oxide), PEO [46]. In this case, water molecules are immobilized while they dissolve PEO crystals. The immobilization during water diffusion through PEO results in apparent non-Fickian behavior. It was observed in block copolymers containing PEO that were studied for water vapor transport that could potentially be used for gas drying or carbon dioxide capture. The morphology of the block copolymer was also found to affect transport in terms of the effective diffusion coefficient, but it did not cause non-Fickian behavior [46, 47].

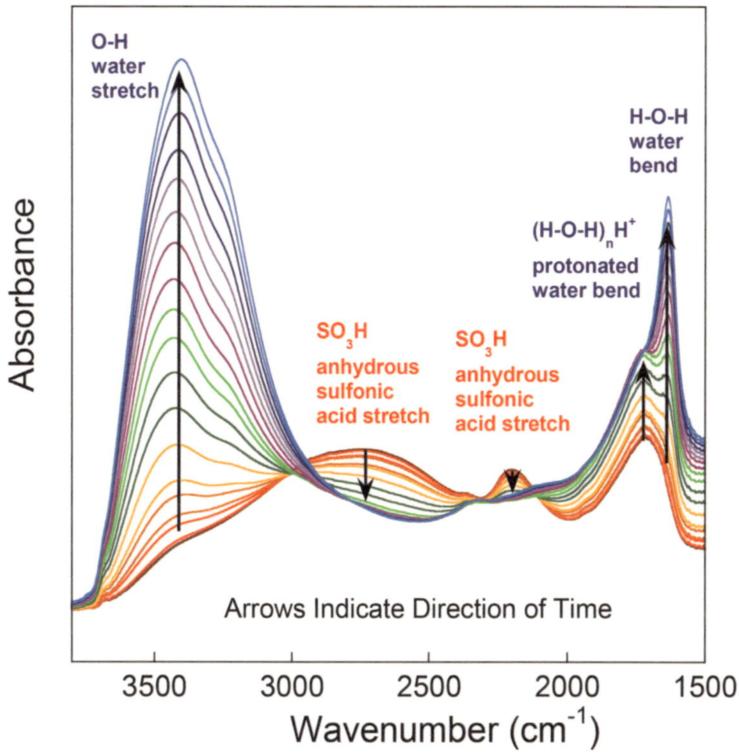

Figure 6.
Experiment of the ***Figure 4(a)*** *type. Dry Nafion® at 30°C is exposed to 100%RH at the top surface, and the FTIR-ATR spectra collected through time. Arrows show the change with time, and the dark red spectrum is time zero (dry Nafion). The equilibrium, hydrated Nafion spectrum is light blue. Reprinted with permission from [20]. Copyright 2010 American Chemical Society.*

4. Conclusions and outlook

In summary, FTIR-ATR is a powerful technique for time-resolved measurements. The specific application to transport in polymer membranes and films has been covered here. It can be used to simultaneously quantify diffusion coefficients, even of multiple species by tracking their distinct FTIR-ATR absorbances [10]. It can also distinguish polymer swelling and other non-diffusive phenomena that occur in tandem with diffusion and cause apparent non-Fickian diffusion. The ability to directly track these non-diffusive processes, such as swelling, reaction, and crystallite dissolution, makes it possible to determine the physical cause of apparent non-Fickian behavior. The ATR set-up itself is quite amenable to transport measurements due to the ability to experimentally define a one-dimensional transport system with well-defined boundary conditions.

A recent development in FTIR-ATR is spectroscopic imaging that has made it possible to glean additional information. It has been used to investigate water and plasticizer distributions in polyvinylchloride membranes [48]. Another interesting approach is the coupling of photoacoustic spectroscopy with FTIR (FTIR-PAS), in order to control the sampling depth. Both FTIR-PAS and FTIR-ATR have been used to study drug transport in a membrane skin model [30].

Other exciting future directions include the growth of FTIR-ATR spectroelectrochemistry, by introducing electrodes into the set-up, which enables electric fields to be applied to polymer electrolytes and desalination membranes for example. The most common use of FTIR-ATR spectroelectrochemistry has been to study the electrochemical doping of electronically conductive polymers that can be electrochemically synthesized directly on the ATR crystal or an intervening electrode [49–56]. There is no reason that this technique could not be extended to other types of membranes, including those that cannot be electrochemically synthesized, and to the study of electric-field-driven transport with or without redox reactions.

Surface-enhanced spectroelectrochemistry [57] is also exciting in that surface specific measurements are possible (much shorter sampling length than the depth of penetration from total internal reflection in a purely dielectric ATR set-up). This enables surface-specific phenomena to be examined and much thinner films to be studied while maintaining the thick film approximation. Another advancement in the study of thin films is polarization modulation infrared reflection absorption spectroscopy (PM-IRRAS) that was applied to water transport in Nafion [58]. This is not an ATR technique, but it is well-suited for studying transport in thin films. Traditional FTIR-ATR has also been applied to thin films, but not for studying transport. It was found that the FTIR-ATR intensity scaled linearly with thickness for sub-micron thick membranes [59].

FTIR-ATR spectroscopy is a versatile, powerful technique for studying multicomponent transport in thin and thick membranes. However, it has been applied to a rather limited set of material systems. Thus, its greatest promise from the membrane transport perspective is in the application to new material systems.

Acknowledgements

This chapter was written with funding support from the National Science Foundation, award numbers 1751450 and 1804871. The author wishes to acknowledge Ashley David for supplying example FTIR-ATR spectra and data.

Conflict of interest

The author declares no conflict of interest.

Author details

Daniel T. Hallinan Jr
Florida A&M University-Florida State University College of Engineering, Chemical and Biomedical Engineering Department and Aero-Propulsion, Mechatronics, and Energy Center, Tallahassee, FL, USA

*Address all correspondence to: dhallinan@eng.famu.fsu.edu

References

[1] Silverstein RM, Kiemle DJ, Webster FX. Spectrometric Identification of Organic Compounds. 7th ed. Hoboken, NJ: John Wiley & Sons; 2005. 502 p. ISBN: 9780471393627

[2] Harrick NJ. Internal Reflection Spectroscopy. New York: Harrick Scientific Corporation; 1979. ISBN: 9780933946132

[3] Hallinan DT Jr. Transport in Polymer Electrolyte Membranes Using Time-Resolved FTIR-ATR Spectroscopy. Philadelphia: Drexel University; 2009

[4] Brandrup J, Immergut EH, editors. Polymer Handbook. 3rd ed. New York: Wiley; 1989. ISBN: 0-471-81244-7

[5] Tan CZ. Temperature-dependent refractive index and material dispersion of diamond. Applied Physics A. 2022; **128**(5):412. DOI: 10.1007/s00339-022-05546-5

[6] Edwards DF, Ochoa E. Infrared refractive index of diamond. Journal of the Optical Society of America. 1981; **71**(5):607-608. DOI: 10.1364/JOSA.71.000607

[7] Li HH. Refractive index of silicon and germanium and its wavelength and temperature derivatives. Journal of Physical and Chemical Reference Data. 1980;**9**(3):561-658. DOI: 10.1063/1.555624

[8] Elabd YA, Baschetti MG, Barbari TA. Time-resolved Fourier transform infrared/attenuated total reflection spectroscopy for the measurement of molecular diffusion in polymers. Journal of Polymer Science Part B-Polymer Physics. 2003;**41**(22):2794-2807. DOI: 10.1002/polb.10661

[9] Jaber JA, Schlenoff JB. Polyelectrolyte multilayers with reversible thermal responsivity. Macromolecules. 2005; **38**(4):1300-1306. DOI: 10.1021/ma0485235

[10] Hallinan DT Jr, Elabd YA. Diffusion and sorption of methanol and water in Nafion using time-resolved Fourier transform infrared-attenuated total reflectance spectroscopy. Journal of Physical Chemistry B. 2007;**111**(46): 13221-13230. DOI: 10.1021/jp075178n

[11] Santos MC, Bendiksen B, Elabd YA. Diffusion of liquid water in free-standing polymer films using pressure-contact time-resolved Fourier transform infrared attenuated total reflectance spectroscopy. Industrial & Engineering Chemistry Research. 2017;**56**(12): 3464-3476. DOI: 10.1021/acs.iecr.7b00114

[12] Fieldson GT, Barbari TA. The use of FTIR-ATR spectroscopy to characterize penetrant diffusion in polymers. Polymer. 1993;**34**(6):1146-1153. DOI: 0032-3861/93/061146-08

[13] Paul DR. Reformulation of the solution-diffusion theory of reverse osmosis. Journal of Membrane Science. 2004;**241**(2):371-386. DOI: 10.1016/j.memsci.2004.05.026

[14] Vasconcelos KL, Bhasin A, Little DN. Measurement of water diffusion in asphalt binders using Fourier transform infrared-attenuated total reflectance. Transportation Research Record. 2010; **2179**:29-38. DOI: 10.3141/2179-04

[15] Lindfors T, Sundfors F, Höfler L, Gyurcsányi RE. FTIR-ATR study of water uptake and diffusion through ion-selective membranes based on plasticized poly(vinyl chloride).

Electroanalysis. 2009;**21**(17–18): 1914-1922. DOI: 10.1002/elan.200904609

[16] Seraji HR, Karimi M, Mahmoudi L. In situ monitoring the change of mechanical response induced by the diffusion of saline water in glassy cellulose acetate. Desalination. 2017;**420**: 191-207. DOI: 10.1016/j.desal.2017.07.013

[17] Sammon C, Yarwood J, Everall N. A FTIR–ATR study of liquid diffusion processes in PET films: Comparison of water with simple alcohols. Polymer. 2000;**41**(7):2521-2534. DOI: 10.1016/S0032-3861(99)00405-X

[18] Davis EM, Theryo G, Hillmyer MA, Cairncross RA, Elabd YA. Liquid water transport in polylactide homo and graft copolymers. ACS Applied Materials & Interfaces. 2011;**3**(10):3997-4006. DOI: 10.1021/am2008618

[19] Mountz DA, Storey RF, Mauritz KA. Fourier transform infrared/attenuated total reflectance analysis of water diffusion in poly styrene-b-isobutylene-b-styrene block copolymer membranes. Journal of Polymer Science Part B-Polymer Physics. 2005;**43**(7):764-776. DOI: 10.1002/polb.20364

[20] Hallinan DT Jr, De Angelis MG, Baschetti MG, Sarti GC, Elabd YA. Non-Fickian diffusion of water in Nafion. Macromolecules. 2010;**43**(10): 4667-4678. DOI: 10.1021/ma100047z

[21] Hajatdoost S, Sammon C, Yarwood J. FTIR–ATR studies of diffusion and perturbation of water in polyelectrolyte thin films. Part 4. Diffusion, perturbation and swelling processes for ionic solutions in SPEES/PES membranes. Polymer. 2002;**43**(6): 1821-1827. DOI: 10.1016/S0032-3861(01)00765-0

[22] Philippe L, Sammon C, Lyon SB, Yarwood J. An FTIR/ATR in situ study of sorption and transport in corrosion protective organic coatings: 1. Water sorption and the role of inhibitor anions. Progress in Organic Coatings. 2004; **49**(4):302-314. DOI: 10.1016/j.porgcoat.2003.07.002

[23] Sundfors F, Höfler L, Gyurcsányi RE, Lindfors T. Influence of poly(3-octylthiophene) on the water transport through methacrylic-acrylic based polymer membranes. Electroanalysis. 2011;**23**(8):1769-1772. DOI: 10.1002/elan.201100076

[24] Sundfors F, Lindfors T, Höfler L, Bereczki R, Gyurcsányi RE. FTIR-ATR study of water uptake and diffusion through ion-selective membranes based on poly(acrylates) and silicone rubber. Analytical Chemistry. 2009;**81**(14): 5925-5934. DOI: 10.1021/ac900727w

[25] Fieldson GT, Barbari TA. Analysis of diffusion in polymers using evanescent field spectroscopy. AIChE Journal. 1995; **41**(4):795-804. DOI: 10.1002/aic.690410406

[26] Beckingham BS, Lynd NA, Miller DJ. Monitoring multicomponent transport using in situ ATR FTIR spectroscopy. Journal of Membrane Science. 2018;**550**: 348-356. DOI: 10.1016/j.memsci.2017.12.072

[27] Carter BM, Dobyns BM, Beckingham BS, Miller DJ. Multicomponent transport of alcohols in an anion exchange membrane measured by in-situ ATR FTIR spectroscopy. Polymer. 2017;**123**:144-152. DOI: 10.1016/j.polymer.2017.06.070

[28] Dobyns BM, Kim JM, Li J, Jiang Z, Beckingham BS. Multicomponent transport of alcohols in Nafion 117 measured by in situ ATR FTIR

spectroscopy. Polymer. 2020;**209**: 123046. DOI: 10.1016/j.polymer.2020.123046

[29] Guo JC, Barbari TA. A dual mode, local equilibrium relaxation model for small molecule diffusion in a glassy polymer. Macromolecules. 2008;**41**(1): 238-245. DOI: 10.1021/ma071662c

[30] Christ A, Szurkowski J, Hanh BD, Wartewig S, Kopycinska M, Neubert RHH, et al. Drug penetration into a membrane investigated by photoacoustic and FTIR-ATR spectroscopy. Analytical Sciences/ Supplements. 2002;**17icpp**:s371-s3s3. DOI: 10.14891/analscisp.17icpp.0.s371.0

[31] Duc Hanh B, Neubert RHH, Wartewig S. Investigation of drug release from suspension using FTIR-ATR technique: Part I. determination of effective diffusion coefficient of drugs. International Journal of Pharmaceutics. 2000;**204**(1):145-150. DOI: 10.1016/ S0378-5173(00)00488-9

[32] Kim K, Hallinan DT. Lithium salt diffusion in diblock copolymer electrolyte using Fourier transform infrared spectroscopy. The Journal of Physical Chemistry B. 2020;**124**(10): 2040-2047. DOI: 10.1021/acs.jpcb.9b11446

[33] Kaim W, Fiedler J. Spectroelectrochemistry: The best of two worlds. Chemical Society Reviews. 2009; **38**(12):3373-3382. DOI: 10.1039/ b504286k

[34] Richey FW, Dyatkin B, Gogotsi Y, Elabd YA. Ion dynamics in porous carbon electrodes in supercapacitors using in situ infrared spectroelectrochemistry. Journal of the American Chemical Society. 2013; **135**(34):12818-12826. DOI: 10.1021/ ja406120e

[35] Newman J, Thomas-Alyea KE. Electrochemical Systems. 3rd ed. NJ Wiley-Interscience: Hoboken; 2004. 647 p. ISBN: 0-471-47756-7

[36] Bard AJ, Faulkner LR. Electrochemical Methods, Fundamentals and Applications. New York: John Wiley & Sons; 2001

[37] Deen WM. Analysis of Transport Phenomena. 2nd ed. Oxford University Press; 2011. 688 p. ISBN: 9780199740284

[38] Silverman M, Hallinan D. The relationship between self-diffusion activation energy and Soret coefficient in binary liquid mixtures. Chemical Engineering Science. 2021;**240**:116660. DOI: 10.1016/j.ces.2021.116660

[39] Diffusion in Polymers. London and New York: Academic Press; 1968. p. 452

[40] Neogi P. Transport phenomena in polymer membranes. In: Neogi P, editor. Diffusion in Polymers. New York: Marcel Dekker; 1996. pp. 173-210

[41] Baschetti MG, Piccinini E, Barbari TA, Sarti GC. Quantitative analysis of polymer dilation during sorption using FTIR-ATR spectroscopy. Macromolecules. 2003;**36**(25): 9574-9584. DOI: 10.1021/ma0302457

[42] Kim K. Dissociation and Transport of Lithium Ions in Polymer Electrolytes. Tallahassee, FL: Florida State University; 2021

[43] Crank J. The Mathematics of Diffusion. 2nd ed. Oxford: Clarendon Press; 1975. ISBN: 0-19-853411-6

[44] Sarti GC. Solvent osmotic stresses and the prediction of case II transport kinetics. Polymer. 1979;**20**(July):827-832

[45] Berens AR, Hopfenberg HB. Diffusion and relaxation in glassy polymer powders: 2. Separation of diffusion and relaxation parameters. Polymer. 1978;**19**(May):489-496

[46] Oparaji O, Zuo X, Hallinan DT Jr. Crystallite dissolution in PEO-based polymers induced by water sorption. Polymer. 2016;**100**:206-218. DOI: 10.1016/j.polymer.2016.08.026

[47] Minelli M, Baschetti MG, Hallinan DT Jr, Balsara NP. Study of gas permeabilities through polystyrene-block-poly(ethylene oxide) copolymers. Journal of Membrane Science. 2013;**432**: 83-89. DOI: 10.1016/j.memsci.2012.12.038

[48] Solovyeva EV, Lu H, Khripoun GA, Mikhelson KN, Kazarian SG. In situ ATR-FTIR spectroscopic imaging of PVC, plasticizer and water in solvent-polymeric ion-selective membrane containing Cd^{2+}-selective neutral ionophore. Journal of Membrane Science. 2021;**619**:118798. DOI: 10.1016/j.memsci.2020.118798

[49] Neugebauer H, Cravino A, Luzzati S, Catellani M, Petr A, Dunsch L, et al. Spectral signatures of positive and negative charged states in doped and photoexcited low band-gap polydithienothiophenes. Synthetic Metals. 2003;**139**(3):747-750. DOI: 10.1016/s0379-6779(03)00292-3

[50] Damlin P, Kvamstrom C, Nyback A, Kaldstrom M, Ivaska A. Electrochemical and spectroelectrochemical study on bilayer films composed of C-60 and poly(3,4-ethylenedioxythiophene) PEDOT. Electrochimica Acta. 2006;**51**(27): 6060-6068. DOI: 10.1016/j.electacta.2006.01.063

[51] Latonen RM, Lonnqvist JE, Jalander L, Kvarnstrom C, Ivaska A. In situ spectroelectrochemical study on a copolymer made from the 2-biphenyl-3-octylthiophene monomer. Electrochimica Acta. 2006;**51**(7): 1244-1254. DOI: 10.1016/j.electacta.2005.06.015

[52] Meana-Esteban B, Kvarnstrom C, Ivaska A. Spectroelectrochemical study on the charging reactions of electrochemically synthesized films from 2-methoxynaphthalene. Synthetic Metals. 2006;**156**(5–6):426-432. DOI: 10.1016/j.synthmet.2006.01.006

[53] Sarac AS, Dogru E, Ates M, Parlak EA. Electrochemical synthesis of N-methylpyrrole and N-methylcarbazole copolymer on carbon fiber microelectrodes, and their characterization. Turkish Journal of Chemistry. 2006;**30**(4):401-418

[54] Osterholm A, Esteban BM, Kvarnstrom C, Ivaska A. Spectroelectrochemical study of the redox reactions of polyazulene on aluminum substrates. Journal of Electroanalytical Chemistry. 2008; **613**(2):160-170. DOI: 10.1016/j.jelechem.2007.10.022

[55] Meana-Esteban B, Balan A, Baran D, Neugebauer H, Toppare L, Sariciftci NS. In situ spectroelectrochemical study of positively and negatively charged states in a donor/acceptor EDOT/benzotriazole-based polymer. Macromolecular Chemistry and Physics. 2011;**212**(22):2459-2466. DOI: 10.1002/macp.201100322

[56] Lepicka K, Pieta P, Shkurenko A, Borawicz P, Majewska M, Rosenkranz M, et al. Spectroelectrochemical approaches to mechanistic aspects of charge transport in meso-nickel(II) Schiff Base electrochromic polymer. Journal of Physical Chemistry C. 2017;**121**(31):

16710-16720. DOI: 10.1021/acs.jpcc.7b04700

[57] Osawa M. Surface-enhanced infrared absorption spectroscopy. In: Chalmers JM, Griffiths PR, editors. Handbook of Vibrational Spectroscopy. Wiley; 2006. DOI: 10.1002/0470027320.s0603 [Online]

[58] Davis EM, Stafford CM, Page KA. Elucidating water transport mechanisms in Nafion thin films. ACS Macro Letters. 2014;**3**(10):1029-1035. DOI: 10.1021/mz500515b

[59] Drazevic E, Kosutic K, Freger V. Permeability and selectivity of reverse osmosis membranes: Correlation to swelling revisited. Water Research. 2014;**49**:444-452. DOI: 10.1016/j.watres.2013.10.029

Chapter 3

Continuous Scan and Repetitive Mode FT-IR Spectroscopy and Its Application in Isomeric Identification, Conformational Analysis and Photochemistry

Prasanta Das

Abstract

This chapter intends to cover the instrumentation of gas phase Fourier transform infrared spectroscopy (FT-IR), its recent advancement, and applications. The major focus have been given to the principle and data acquisition scheme of the repetitive mode measurement method of FT-IR spectrometer. The application of this spectroscopy in the isomeric identification of the methylated polycyclic aromatic hydrocarbons (MPAHs) and the conformational analysis of diols have been discussed. Furthermore, the application of the repetitive measurement mode of FT-IR combined with the UV laser in monitoring the atmospherically relevant photochemical reactions has been covered. In conclusion, this chapter briefly summarizes the current applications and discusses future applications of this technique in following drug degradation.

Keywords: continuous scan FT-IR, repetitive mode time-resolved FT-IR, isomeric identification, conformational analysis, atmospheric photochemistry, drug degradation

1. Introduction

Fourier transform infrared spectroscopy (FT-IR) is an analytical tool that is used to obtain molecular structural information based on spectral signatures of molecular vibrations. It has been employed in numerous research fields such as chemistry, chemical industries, chemical engineering, pharmaceutical industries, atmospheric chemistry, etc. Thus, it becomes a standard instrument in modern chemical and physical laboratories.

In a conventional FT-IR spectrometer, the movable mirror of the Michelson interferometer undergoes a complete movement to produce the interferogram which is Fourier transformed to obtain a spectrum. This technique is used for getting vibrational information of molecules and molecular complexes which help in identification of isomers and conformers and finding molecular interactions such as hydrogen bonding, etc. It takes a certain amount of time for the complete movement of the mirror and thus restricts the application of FT-IR to time-evolving process.

Since its discovery, it has been passed through several modifications especially to reduce the data acquisition time so that molecular changes that happened during a reaction can be investigated in-situ [1]. These modifications include the development of time-resolved FT-IR spectroscopy such as repetitive mode [2], rapid scan [3, 4], and step-scan [3, 5, 6] which covers, typically, time-resolution of 10^2–10^{-9} s. Several commercial FT-IR spectrometers are equipped with these options. In the case of repetitive mode measurement, several spectra are recorded in a conventional manner at a particular time interval with the continuous light illumination for photolysis; a typical time resolution that can be obtained with this technique is ~100 ms. In the case of rapid scan mode [6], it is possible to move the mirror at high speed to record the spectra in a short time of 10's ms. It enables to record a series of spectra after laser flashes in a time-resolved way. The series of time-resolved difference spectra can be obtained by subtracting the spectrum recorded before laser excitation. In the step-scan FT-IR spectroscopy [6], data acquisition occurs in multiple steps of the movable mirror. At each step, a laser is fired to induce the reaction, and then the IR transient induced by the laser flash is recorded. Interferograms and therefore spectra are reconstructed from a series of transients corresponding to a complete series of mirror stop position [4, 6]. In the last two decays, the step-scan FT-IR spectrometer has been coupled with the imaging technique and applied in the area of biomedical science [7–9].

The infrared spectra of molecules can be measured in various states of materials - solid, liquid, and gas phases. The former two states' measurements are common and used in routine studies, but the later mode measurement is uncommon since the experimental requirement is more for acquiring the gas phase spectra. Furthermore, in some cases, it is way difficult to get enough molecules into the gas phase, *e.g.*, polycyclic aromatic hydrocarbons of more than four member rings have shown vapor pressure below 10^{-6} mmHg at room temperature. However, among these three states, the gas phase spectra provide more valuable information. This is illustrated by comparing spectra of 1,4-butanediol ($HO\text{-}C_4H_8\text{-}OH$) in the gas phase [10] and liquid state [11], see **Figure 1a** and **b**.

From **Figure 1**, we can see that above 3500 cm^{-1} there are three IR absorption bands in the gas phase spectra of 1,4-butanediol. Among these, the band of higher

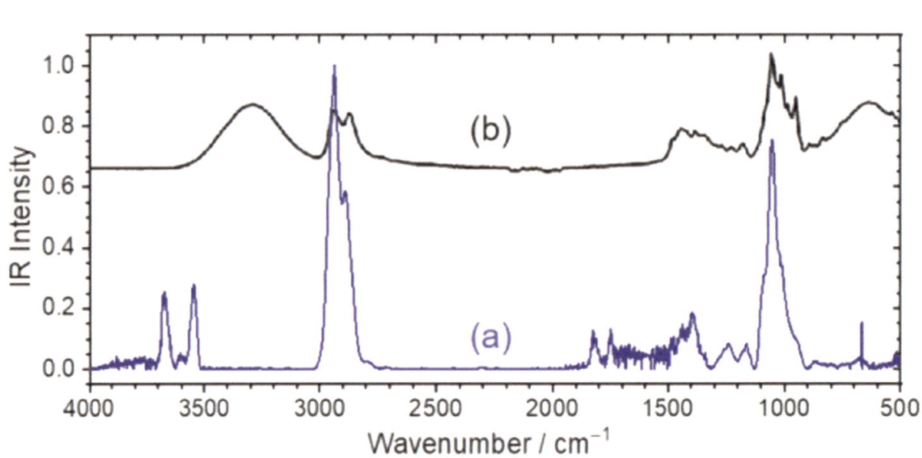

Figure 1.
Infrared spectra of 1,4-butandiol in the gas phase (a) and liquid state(b). Spectra (b) converted from % transmission mode to absorbance mode. Spectra (a) reproduced from reference [10] with permission; Copyright (2015) ACS. Spectra (b) reproduced from reference [11] with permission; Copyright (2017) RSC.

frequency one is assigned to the free O-H stretching vibration, and the remaining two bands of lower frequency value are assigned to the hydrogen-bonded O-H stretching vibrations which are belonging to the two different conformers [10]. These information are hidden in the corresponding liquid state spectra, only a broad IR absorption band appeared at $\sim$3300 cm^{-1}; this is what you can find in most textbooks. The other methods of sampling techniques along with instrumentation have been summarized by Marwa El-Azazy [12].

This chapter means to give an overview of the following. From the above discussion, it is clear that for measuring infrared spectra of low vapor pressure compounds, one can use a long-path gas cell to improve the detection limit. Therefore, how the FT-IR spectrometer can be assembled with the gas cell and vacuum lines to investigate the vibrational spectra of molecules and molecular complexes in the gas phase. In addition, if this basic set-up is modified and coupled with the UV laser whether that can be used to monitor photochemical reactions. Furthermore, how, it can be used to obtain quantitative information on free molecules, molecules in a mixture, and reactants and photoproducts in a reaction. Discuss the applications of these methods in the gas phase vibrational spectroscopy of methylated polycyclic aromatic hydrocarbon (MPAHs), conformation analysis of diols, and monitoring the photochemical reactions of halobenzenes.

2. Instrumentation

The principle and instrumentation of FT-IR spectrometer can be found in many textbooks, including the book written by Griffiths and Haseth [13]. Nevertheless, a brief description is required to make the connectivity. The design of interferometers used for IR spectroscopy is based on the interferometer which was originally designed in 1891 by Michelson [14]. This interferometer is a device that split a beam of radiation into two beams and then recombines the two beams after a path difference has been introduced. A condition thereby created under which interference between two beams can occur. The variation of intensity of the beam emerging from the interferometer is measured as a function of the path difference called retardation which is designated as δ. The schematic diagram of Michelson interferometer, interferogram, and spectrum for monochromatic light source is shown in **Figure 2** [15]. Here a constructive interference will occur only when the distance between the movable mirror and the fixed mirror (*i.e.,* δ) is an integral multiple of λ, whereas a destructive interference will take place when δ is an integral multiple of $\lambda/2$.

To obtain an interferogram, $I(\delta)$, the detector signal is digitized and recorded as a function of retardation (δ). A reference He-Ne laser is employed to monitor movement of movable mirror. The interferogram intensity of polychromatic source is mathematically described by Eq. (1):

$$I(\delta) = \int_{-\infty}^{+\infty} B(\nu) \cos(2\pi\nu\delta) d\nu \tag{1}$$

where $B(\nu)$ is the spectral intensity at wave number ν (in cm^{-1}). The Fourier transformation (FT) of $I(\delta)$ gives the single-beam IR spectrum expressed by Eq. (2):

$$B(\nu) = \int_{-\infty}^{+\infty} I(\delta) \cos(2\pi\nu\delta) d\delta \tag{2}$$

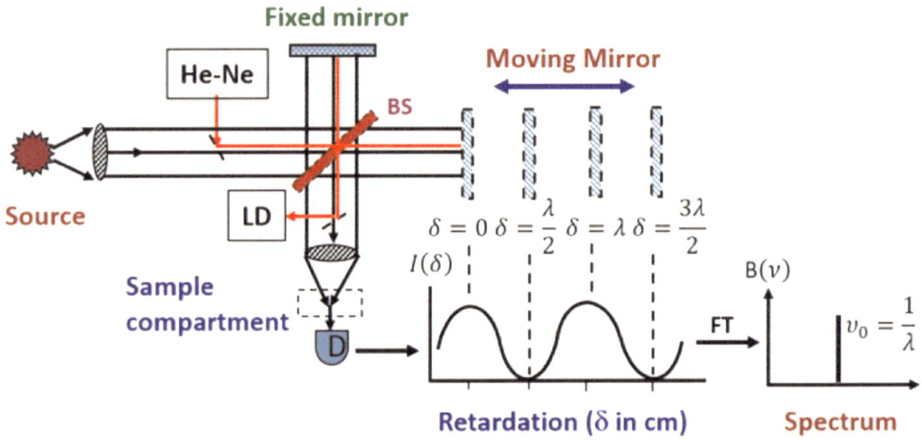

Figure 2.
Optical diagram of a Michelson interferometer with the He-Ne laser running co-axial, interferogram and spectrum for a single frequency light source. The interferometer consists of four major components: A fixed mirror, a moving mirror, BS: Beamsplitter, D: Detector. LD: He-Ne laser detector. Adapted from reference [15]. Copyright (2011) Indian Institute of Science Bangalore (https://etd.iisc.ac.in/handle/2005/1252).

The movable mirror moves continuously in case of continuous scan mode as discuss below.

2.1 Continuous scan interferometry

The continuous-scan is preferably used for the routine static or very slow kinetic measurements. In a continuous-scan FT-IR spectrometer, the moving mirror moves continuously at a constant velocity, $V(\text{cm s}^{-1})$, and the optical path difference at time $t(\text{s})$ is given by $\delta = 2Vt(cm)$. The interferogram data points are digitized at the zero crossings of the He-Ne laser signal as shown in **Figure 3** [15]. The use of laser signal ensures that $I(\delta)$ is measured at precisely equal intervals of mirror positions and provides an internal wavelength calibration for every scan. Because of continuous

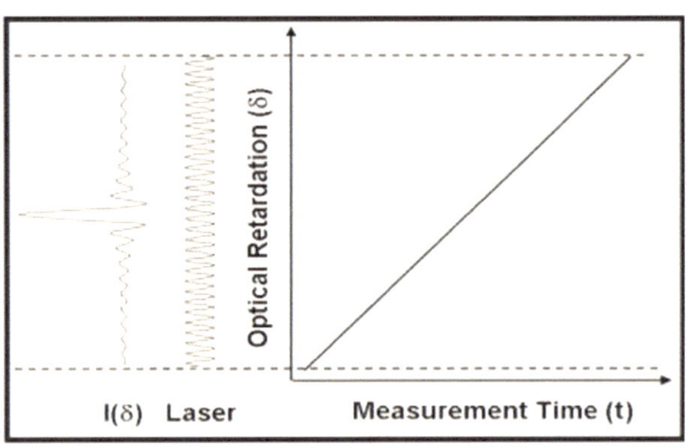

Figure 3.
Schematic of data collection in continuous-scan interferometer. Adapted from reference [15]. Copyright (2011) Indian Institute of Science Bangalore (https://etd.iisc.ac.in/handle/2005/1252).

movement of the mirror, the interferogram $I(\delta)$ changes as a function of time. The Fourier frequency (f_F *in Hz*) of IR light at a wave number ν (in cm^{-1}) is given by:

$$f_F = 2V\nu \tag{3}$$

where V is the mirror velocity in cm s^{-1}. Typically, the slow velocities are chosen for thermal detectors; such as deuterated triglycine sulfate (DTGS), whereas, the fast velocities are chosen for fast detectors, such as a mercury cadmium telluride (MCT) or indium antimonide (InSb) detector, for routine or kinetic measurements.

2.2 Repetitive measurements with FT-IR

The modern FT-IR spectrometer has a provision for the repetitive mode measurement. The data acquisition scheme is similar to normal scan [I(δ) vs δ] which is repeated for many times depending on the requirement to follow up the reaction kinetics completely. Thereby introduced another dimension in in conventional scan mode, and thus it is a three dimensional [I(δ) vs δ and t]. **Figure 4** shows the data collection scheme of repetitive mode measurement of the FT-IR. For a set of experiment, say, five interferogram (t1 - t5) correspond to the five spectra at time resolution $\Delta t = (t_2 - t_1)$ and each interferogram is an average over many continuous scans as describe in **Figure 3**.

For a particular spectral resolution, the time resolution is limited by scanning speed of movable mirror and number of co-additions, *i.e.*, no of scans consider for the spectra. This kind of measurements can be used to monitor photochemical and thermally perturbed process without any break. The advantage of this mode measurement

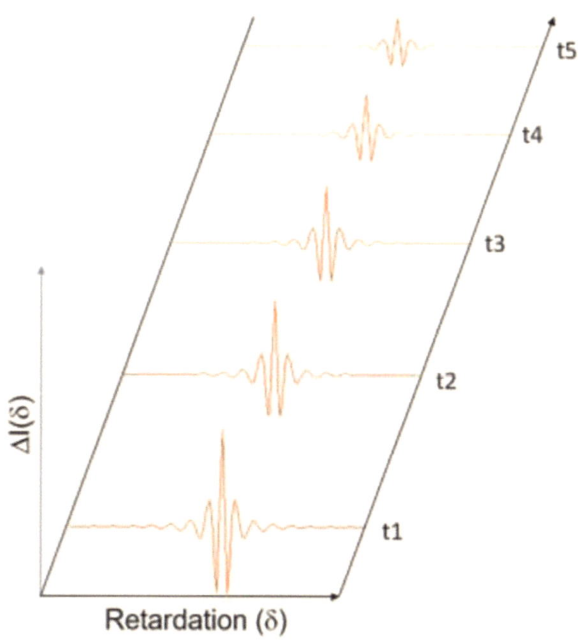

Figure 4.
Schematic of data collection in repetitive mode of measurements. Interferogram t1 - t5 corresponding to the spectra 1–5, respectively. Each interferogram or thereby spectra correspond to the average over many conventional spectrums.

is that one can do photolysis and simultaneously keep recording the spectra of sample which is not refresh like in the case molecular beam technique. That means it has capability to monitor photolysis products on-line. This is like a method which can be used to study the photo- and thermo-chemical evolution within a well-defined system. Thus, this methodology is well suited for applied problems.

2.3 Variable path-length long-path gas cell

For measuring infrared spectra of molecules and/or monitoring a photochemical reaction in the gas phase, the well-known method is to couple the gas cell with the FT-IR spectrometer. However, in some cases it is difficult to obtain sufficient amount of compounds into the gas phase from its solid or liquid states for the IR spectroscopic study using a normal gas cell of ~10 cm (typically) optical path length (OPL). Also, this is true in cases where photolysis product yield is found to be low. In such cases, where concentrations are below the detection limit with the gas cell, the common strategy is to use longer OPL (l) to improve the IR absorbance. This is in accordance with the Lambert–Beer's law. In other words, a higher detection limit can be achieved by employing a long-path gas cell.

Figure 5a shows a schematic of the variable path-length long-path gas cell which was made according to the "white cell" principle [16]. The cell contains three internal gold-coated mirrors; two (M1 and M2) of them are adjustable attached at the top and one (M3) is fixed at the bottom of the cell called a field mirror. These adjustable mirrors are placed at the radius of curvature of the fixed mirror. At the entrance and exit point of the IR radiation two ZnSe windows are used. Outside of both the windows two adjustable transfer mirrors are used to guide the IR light from the source to the detector through the gas cell.

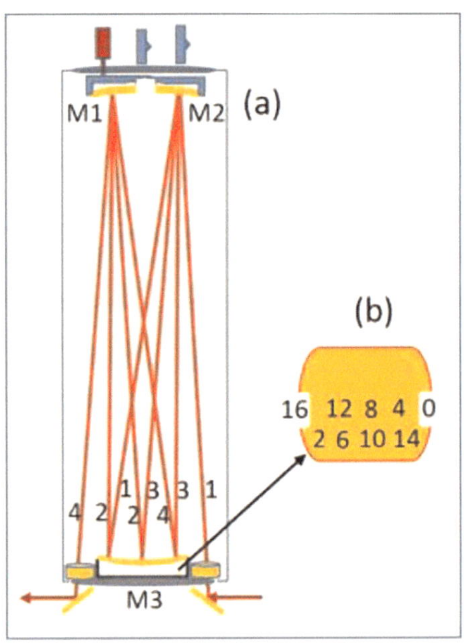

Figure 5.
(a) Schematic diagram of typical "white cell" [side view]; M1, M2, M3 - mirrors, numbers (1–4) indicate the direction of light. (b) Top view of field mirror (M3); representative light images or spots on this mirror are indicated by numbers.

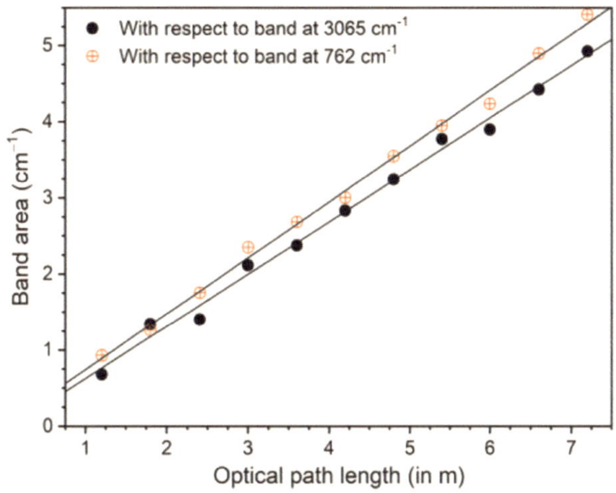

Figure 6.
Observed band area (in cm^{-1}) vs optical path length (in m) for naphthalene. Adapted from reference [15] with permission. Copyright (2011) Indian Institute of Science, Bangalore (https://etd.iisc.ac.in/handle/2005/1252).

The OPL of the cell can be varied by changing the position of the adjustable mirror. The beam comes through one window (right-hand side) and reflects back and forth between the field mirror and the adjustable mirrors of the gas cell. After every 4 passes, the beam migrates across top of the field mirror on the input side until it misses that mirror and exits through the exit window. The field mirror shows the migration of the IR beam in the gas cell. The IR beam comes in at the position marked 0 and exits at the position marked 16, see **Figure 5b**.

Initially, the desired OPL is set roughly by adjusting the position of the adjustable mirror with the help of red knob in **Figure 5a** and simultaneously counting the He-Ne laser spots on the fixed mirror. From the number of spots as recommended by manufacturer, the IR light OPL can be calculated back. Further fine-tuning of OPL can be done by adjusting mirror slightly forward or backward and looking for the maximum interferogram signal with the empty cell. Before using the multi-pass gas cell for quantitative measurements, one needs to make sure optical path lengths change precisely and follow the linearity in absorbance *vs* OPL plot. **Figure 6** shows the results of such calibration curve obtained using naphthalene as a standard compound. This has been carried out at room temperature. A similar calibration method is applied while setting the precursor amount for the photo-chemical reaction studies.

2.4 Experimental set-up for the gas phase studies

The description of the experimental set-up can be found in Ref.s [17–20]. **Figure 7** shows how a variable multi-pass long-path gas cell was coupled with the FT-IR spectrometer and vacuum lines. FT-IR spectrometer utilized in this set-up was equipped with a liquid-nitrogen-cooled photovoltaic mercury-cadmium-telluride (PV-LN-MCT) detector and a KBr beamsplitter. The vacuum line can be solely homemade using 1/4″ valves, tubing, and fittings from Swagelok. It is connected to the gas cell, pressure transducer, vacuum pump, Ultra-Torr for the sample tube and bulb, and a gas reservoir cylinder. The sample (solid/liquid) holder

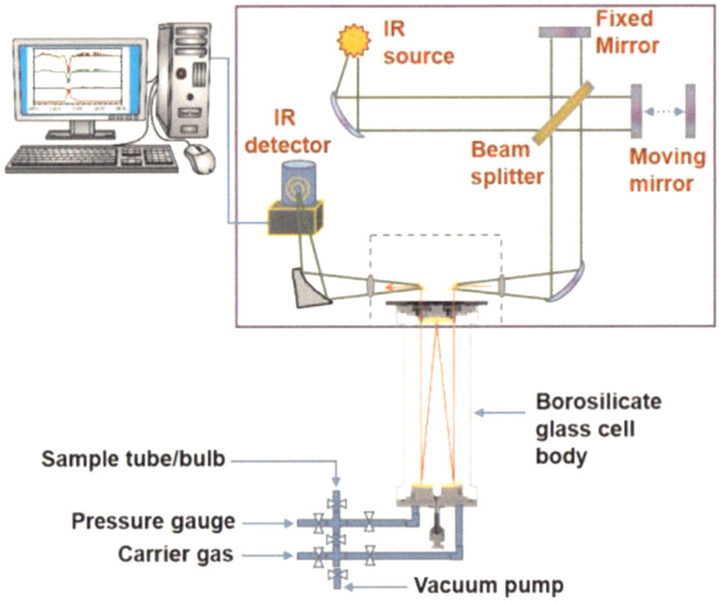

Figure 7.
Schematic diagram of FT-IR spectrometer coupled with a variable multi-pass long-path gas cell and vacuum lines.

is a glass tube/bulb which is attached to the vacuum line using Ultra-Torr fitting and O-ring.

The sample can be loaded in a glass tube or bulb depending on the state of the sample. The dissolved gases are removed by applying several cycles of freeze-pump-thaw. The sample vapor was first allowed to expand into the vacuum line and then transfer to the gas cell and mixed with carrier gas argon (Ar, UHP). This is done in cases where solid or liquid samples have sufficient vapor pressure at room temperature but need to have a temperature controlling system to raise the T during sampling and to maintain cell T during FT-IR measurements for samples having low vapor pressure at room. The spectrometer and sample compartment were continuously purged with ultra-high purity (UHP) nitrogen gas during experiments to avoid interference from moisture and carbon dioxide. Argon (Ar) gas kept in the reservoir cylinder is used for cleaning the cell as well as carrier gas.

This setup was used for investigating vibrational spectra of gaseous PAHs and MPAHs. The vapor pressure of such compounds is low at room temperature, ranging from 10^{-2} to 10^{-7} mmHg for 2 to 7 rings system [15]. Therefore, cell needs to heat up during sampling and acquiring spectra. For very low ($< 10^{-7}$ mmHg) vapor pressure compound, the recommended way is to fill the sample bulb with a solid sample and placed it inside the cell. This setup is also used for the conformational analysis of diols in the gas phase. Furthermore, this basic setup has been modified to investigate the photochemical reactions in the gas phase.

2.5 Experimental set-up for photochemical reaction studies

The repetitive mode measurement method of FT-IR was used to monitor the photochemical reaction in the gas phase. **Figure 8** shows the experimental setup [21]. The gas cell shown in this setup served as a reactor in this experiment. Its borosilicate

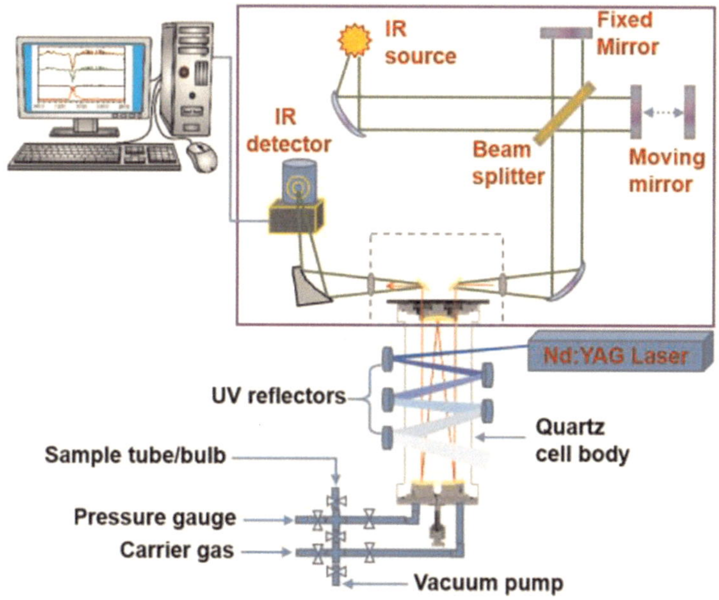

Figure 8.
Schematic diagram of FT-IR spectrometer coupled with the variable multi-pass long-path gas cell, vacuum lines and UV laser (ns).

glass body was replaced with a Quartz tube to allow UV light for the photolysis. The 266 nm light was generated from the fourth harmonic (1064 nm) of a pulse Nd:YAG laser. In order to increase the photolysis efficiency, a multi-pass arrangement was made for the laser light as well using five UV reflectors. In past, a similar approach has been applied by Eberhard et al. [22]. They used a pair of rectangular UV reflectors in order to make a multi-pass arrangement across the fixed gas cell which was coupled with a step-scan time-resolved FT-IR spectrometer. In this set-up, the reflectors were placed on circular holders which were hanging on a rectangular frame. The frame has a provision for "x" and "y" movements [21]. This reflector set-up was placed parallel to the cell body to benefit six times UV light passes through the gas cell.

Usually, the laser pulses energy is measured before entering the gas cell. However, inside the cell, the energy is lower due to absorption by cell body material and UV reflector, which is taken into account in the photolysis efficiency estimation. The amount of precursor was set after measuring its infrared spectra as a function of sample pressure so as to keep the absorbance ~1.2. Typically, a small amount of sample vapor was loaded, as much to keep the IR absorbance value in the working range, in the gas cell for the photolysis studies. The partial pressure of sample inside the gas cell can be estimated from its infrared integrated absorbance as described in **Section 3.2**.

First measured the infrared spectrum of the precursor without photolysis. The precursor stability, as well as its rate of condensation, was tested prior to photolysis by measuring its infrared spectra for up to several minutes. Then the irradiation was started, and simultaneously infrared spectra were recorded. To monitor the reaction, several spectra at different photolysis time intervals have been recorded using the repeated mode measurements of FT-IR. Point to be noted, the laser light irradiation and repetitive scan measurements were started at the same time manually. The precursor spectra were subtracted with proper weighing factors from the spectra recorded during the photolysis to get the product spectra which are called difference spectra. The depletion and formation of infrared bands in the difference spectra indicate the decomposition of precursor and formation of products, respectively.

There is always a trade-off to play with spectral resolution and number of scans per spectra to achieve a maximum temporal resolution. The FT-IR, used in this set-up, required 414 ms per scan at a spectral resolution of 2 cm^{-1}. Therefore, monitoring a photochemical reaction of lifetime 100's of ms with spectral resolution of 2 cm^{-1} is possible with this technique.

There are certain advantages to this technique. It can measure a wide spectral range which allows the detection of many species simultaneously. Furthermore, this technique can be used to study the chemical evolution within a well-defined isolated system; which means, it is well suited for applied problems. The disadvantage of this technique is that it suffers a lack of sensitivity as compared to the technique based on molecular beam combined with laser and mass spectrometry. However, one can, to some extent, improves the sensitivity of this technique by using a long-path gas cell. This method is used to investigate the photodegradation pathways of atmospherically and industrially relevant halogenated compounds [2].

3. Quantitative analysis

3.1 Infrared intensity

The quantity of interest in the intensity calculation is the integrated absorption coefficient A (in km mol^{-1}). It can be determined theoretically using Eq. (4) [23]

$$A_i = 42.254\left|\frac{\partial \mu}{\partial Q_i}\right|^2, \tag{4}$$

where $\partial\mu/\partial Q_i$ are the dipole moment derivatives in D (Å $amu^{1/2}$) evaluated via analytical derivatives computed at the DFT level of theory.

On the other hand, absolute intensity (in cm^{-2} atm^{-1}) can be obtained experimentally by using the Eq. (5) [24]

$$A_i = \frac{2.303\int \log\left(I_o/I\right)_i d\nu_i}{P_i l}, \tag{5}$$

where l (in cm) is the OPL and P_i's are the vapor pressures. It is not possible to get the vapor pressure of PAHs or methyl- and dimethyl-PAHs by conventional pressure gauge system since they are seeded on carrier gas(Ar). Therefore, one can estimate the partial pressures (P_i) under all the bands with the help of the observed band areas and their corresponding calculated intensities by using the same equation. The average pressure obtained from all the bands were then calculated using $\mathbf{P} = \sum_{\mathbf{i}}^{\mathbf{n}} \frac{\mathbf{P_i}}{\mathbf{n}}$ where n is the total number of observed bands which was used to calculate the observed band intensities (in cm^{-2} atm^{-1}). In order to get the experimental intensities in km mol^{-1}, the values in cm^{-2} atm^{-1} were multiplied by a factor of 82.056 (T/K).

3.2 Estimation of reactant and photoproducts

The amount of sample (reactant) decay and formation of new products can be measured from the integrated band area of the observed infrared spectra using Eq. (6) [25], which is basically the modified Eq. (5).

$$P_i(atm) = \frac{2.303 \times 82.05 \times T \times 10^{-5} \times \tau_i}{A_i \times l}, \tag{6}$$

in which, τ_i is the integrated absorbance (in cm^{-1}), A_i (in km mol^{-1}) is the absorption coefficient, l (in cm) is the OPL, and T (298 K) is the experimental temperature. Typically, we chose 2–3 absorption lines of a species and averaged the derived partial pressure. In case of rotational-resolved spectra, sum of band areas over all resolved lines is assumed to be equivalent to the unresolved band area within error limit. The derived pressure errors are typically within ±20% due to the fitting errors.

3.3 Photolysis efficiency

The estimation of the photolysis yield is discussed in this section. It helps to know the efficiency of experimental set-up. In order to do that lets take photolysis of (0.43 ± 0.01) torr C_6H_5Cl at 266 nm as an example. The photolysis decay (x) of precursor (C_6H_5Cl) is estimated using Eq. (7)

$$x = n(C_6H_5Cl) \times \sigma(266nm) \times F \tag{7}$$

where, n is the amount of C_6H_5Cl (2.829×10^{17}) molecules present in the laser active volume 20.3 cm^{-3}, F is the laser fluence 13.181×10^{16} photon cm^{-2} $pulse^{-1}$ [assuming over all 25% loss on measure energy 37 mJ $pulse^{-1}$ due to cell and UV reflectors absorption and laser active area 0.283 cm^2 for ~6 mm laser light beam diameter], and σ is the absorption cross section of C_6H_5Cl (4.933×10^{-19} cm^2 $molecule^{-1}$) which is according to the reported ε(266 nm) = 129 cm^{-1} L mol^{-1} [26]. According to Eq. (1), the amount of photoexcited C_6H_5Cl or decay of it's found to be 1.832×10^{16} molecules $pulse^{-1}$ in 20.3 cm^3. Eventually, this many photoproducts get distributed over the cell volume 1.805×10^3 cm^3 and thus value of x is estimated to be 1.015 x 10^{13} molecules cm^{-3} $pulse^{-1}$ which corresponds to 3.129×10^{-4} torr $pulse^{-1}$. The laser is operated at 10 Hz repetition rate and thus overall accumulated decay of precursor during 53 s photolysis is estimated to be 0.166 torr; that means efficiency is 38.6%. In other words, it's clear that dropping of 0.166 torr C_6H_5Cl is the lower limit for observing major photolysis products during first 53 s. More details of this quantitative analysis method can be found in the forthcoming article [21].

4. Applications

This section is dedicated to discussing a few representative applications of the experimental set-ups presented in **Figures** 7 and 8. In past many system have been investigated using these set-ups: PAHs and MPAHs namely 1- and 2-methylnaphthalenes (MNs) [27]; 1,5-, 1,6-, and 2,6-dimethylnaphthalenes (DMNs) [28]; 2,4-, 2,6-, 2,7-, and 2,8-dimethylquinolines (DMQs) [25]; 1,9-, 2,4-, and

3,9-dimethylphenanthrenes (DMPs) [20]; and fluorene, 1-methylfluorene and 1,8-dimethylfluorne [19]; conformational analysis of diols [10]; hydrogen bonded oligomers of methanol [29]; and photodegradation pathways of halobenzenes [2]. But discussion will be restricted to exemplary a few systems as discussed in the subsequent sections.

4.1 Isomeric identification of methylated-PAHs

Das et al. [28] have reported the gas phase IR spectra of 1,5-DMN, 1,6-DMN, and 2,6-DMN using the technique presented in **Figure 7** and assigned their vibrational modes using density functional theory (DFT) calculations. **Figure 9** shows gas-phase infrared spectra of DMNs. Also, quantitative estimation has done as discussed in **Section 3.1**; vapor pressure of DMNs were found to be 2.1–2.5 $\times$ 10^{-2} mmHg at 90°C.

Table 1 listed four characteristic bands of the DMNs. These bands may help in their identification in an unknown mixture. The aromatic C-H out-of-plane bending vibration is the strongest among all the bands in DMNs and is easy to identify near 800 cm^{-1}. The DMNs are distinguishable from the position of this band which appears with a clear separation in different DMNs. A set of three bands assigned for

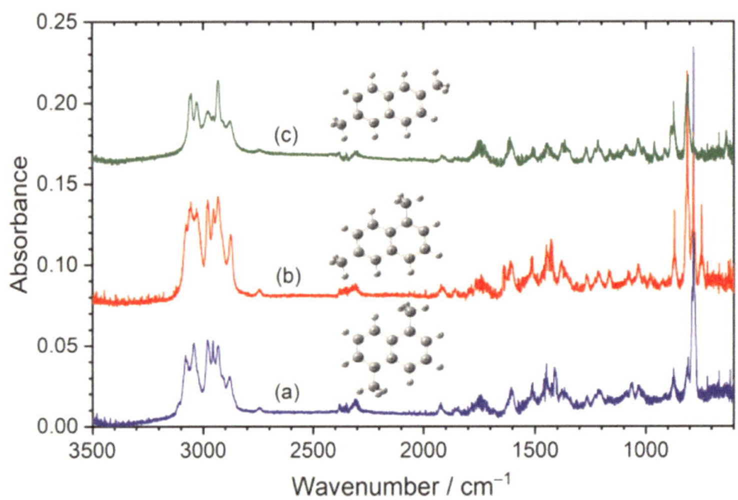

Figure 9.
The infrared absorption spectra for (a) 1,5-DMN; (b) 1,6-DMN; and (c) 2,6-DMN at 0.5 cm^{-1} resolution. Reproduced from reference [28] with permission. Copyright (2008) Elsevier.

Mode of vibration	1,5-DMN[c]	1,6-DMN[c]	2,6-DMN[c]
Aromatic C-H out-of-plane	783.2 (1.000)	812.8 (1.000)	808.5 (1.000)
Methyl C-H stretching	2882.4[a] (0.146) 2931.8[b] (0.107) 2956.9[b] (0.050)	2874.8[a] (0.395) 2933.3[b] (0.395) 2953.8[b] (0.079)	2880.0[a] (0.212) 2933.9[b] (0.549) 2955.2[b] (0.012)

[c]*The band positions are given in cm^{-1} and the relative intensities in parentheses.*
Reproduced from reference [28] with permission. Copyright (2008) Elsevier.

Table 1.
Comparison of observed aromatic (C-H) out-of-plane bending and methyl C-H symmetric[a] and asymmetric stretching[b] vibrations in DMNs.

Mode of vibration	1-MN[a]	2-MN[a]
Aromatic C-H stretching	3076.9 (0.544)	3062.9 (1.771)
Aromatic C-C stretching	1077.2 (0.113)	1644.6 (0.206)
Aromatic C-H in-plane bending	1399.7 (0.181)	1134.6 (0.151)
Aromatic C-H out-of-plane bending	978.8 (0.087) 788.9 (1.000)	952.2 (0.152) 812.1 (1.000)

[a]*Frequency values are in cm^{-1}. In parenthesis, relative intensities are given.*
Reproduced from reference [27] with permission. Copyright (2012) Indian Association for the Cultivation of science.

Table 2.
A few characteristic and unique bands of MNs.

the methyl C-H symmetric and asymmetric stretching is the next set of bands that are distinct in all the DMNs. They appear clearly resolved at the high-frequency region of the IR spectrum near 2900 cm^{-1} with moderate intensities and can be easily marked.

Chakraborty et al. [27] have proposed a similar identification of MNs. A few aromatic C-H stretching and aromatic C-H out-of-plane bending vibrations has been identified in the experimental spectra of MNs as evidenced from reference [27]. Out of these, a few either intense or unique bands have been chosen, as listed in **Table 2**, for the isomeric identification.

In 1- and 2-MN, an intense aromatic C-H stretch has been identified at 3076.9 and 3062.9 cm^{-1}, respectively. This band is separated by 14 cm^{-1} from one isomer to the another. Therefore, this band can be used for the isomeric identification of MN in a complex mixture. One unique band was observed at 1644.6 cm^{-1} in 2-MN and at 1077.2 cm^{-1} in 1-MN. These bands are unique to those isomers and are not observed in the other isomer. Another unique band has been identified in the experimental spectra of 1- and 2-MN at 1399.7 and 1134.6 cm^{-1}, respectively, for the aromatic C-H in-plane bending vibration. This band is of low intensity and clearly visible in the recorded spectra of MNs. There are two sets of bands that are identified for aromatic C-H out-of-plane bending vibrations at 978.8 and 788.9 cm^{-1} in 1-MN and 952.2 and 812.1 cm^{-1} in 2-MN. The first set of bands is highly intense, whereas the second set is of low intensity. These bands are clearly distinguishable for different isomers of MN. Therefore, spectral bands in the 1800–500 cm^{-1} region will be helpful for the isomeric identification of MNs in a complex mixture. After careful inspection of the high-intensity aromatic C-H out-of-plane bending vibrations in the region 1800–500 cm^{-1} and at the aromatic C-H stretching vibrations in the region 3200–2800 cm^{-1}, it is possible to distinguish between the MNs. The isomeric identification through infrared spectra of these compounds, as suggested here, will perhaps be relevant in the field of environmental and atmospheric chemistry.

Similarly, a few characteristic infrared absorptions have been found for other series of methylated PAHs such as DMPs and DMQs. The details assignment and isomeric identifications for these systems are beyond the scope of discussion here. Therefore, interested readers are directed to go through reference [20, 25].

4.2 Conformational analysis using steady state measurements

Das et al. [10] applied the technique presented in **Figure 7** to measure the infrared spectra of 1,2-ethandiol (1,2-ED) and 1,4-butanediol (1,4-BD) in the gas phase at different T's. The corresponding observed spectra are presented

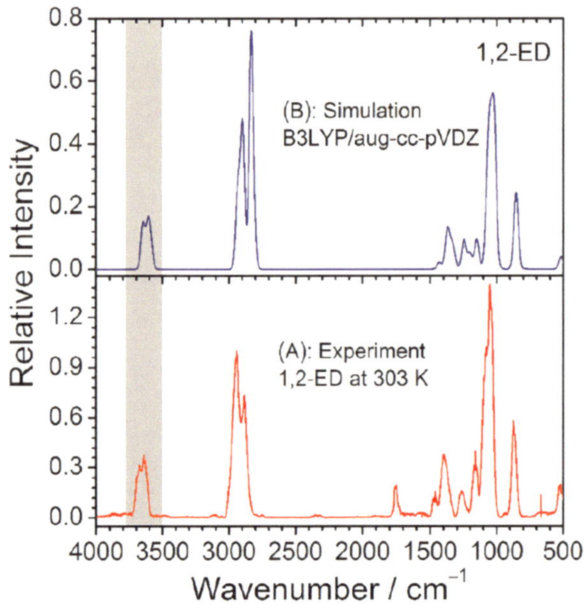

Figure 10.
The experimental IR absorption spectra of 1,2-ED in the gas phase at 303 K (A) and stimulated spectrum of a mixture of stable conformers tGg′, gGg′, g′Gg′, tTt, tTg, gTg′, gTg, gGg, tGt, and tGg (B). B3LYP/aug-cc-pVDZ calculated anharmonic intensities in the individual spectra of conformers were weighted by the population of the respective conformer. The simulated spectrum (B) was obtained using Gaussian functions centered at the anharmonic frequencies and with a bandwidth at half-height of 35 cm^{-1}. The O-H stretching region marked with gray. Reproduced from reference [10] with permission. Copyright (2015) ACS.

in **Figures 10A** and **11A′** and compared with the simulated spectra constructed based on a mixture of conformers, **Figures 10B** and **11B′** and done the conformational analysis.

A statistical thermodynamic population analysis at experimental temperatures have done with the chosen 10 conformers of 1,2-ED (tGg′, gGg′, g′Gg′, tTt, tTg, gTg′, gTg, gGg, tGt, and tGg) and 1,4-BD (g′GG′Gt, gG′G′Gt, tG′TGt, g′TTGt, gGTGt, tTTTt, tGGGt, tTGG′t, gTGGt, and tTGTt). The purpose of the theoretical population analysis is to predict which conformers contribute to the experimentally measured vibrational spectra. The standard statistical mechanical relations were used to calculate the free energies of chosen conformers of 1,2-ED and 1,4-BD using DFT (B3LYP/aug-cc-pVDZ). Then, the fractional gas phase equilibrium population [10, 30–32], P(M) of a conformer M is calculated according to the Boltzmann distribution

$$P(M) = \frac{\mathbf{exp}\left(-\frac{\Delta G_M}{RT}\right)}{\sum_i \mathbf{exp}\left(-\frac{\Delta G_i}{RT}\right)} \quad (8)$$

where i spans all the conformers of 1,2-ED and 1,4-BD, respectively. The calculated $P(M)$ at experimental temperatures were multiplied by 100 in to order to get the percentage population. It was assumed that population of each conformer corresponds to its weight at the experimental temperatures. The stimulated spectrum **Figure 10B** of 1,2-ED is a mixture of conformers tGg′, gGg′, g′Gg′, tTt, tTg, gTg′, gTg, gGg, tGt, and tGg with a contribution of 55.5, 22.6, 13.4, 0.5, 2.8, 1.3, 0.4, 1.0, 1.0, and 1.5%, respectively. Similarly, the populations estimated with the

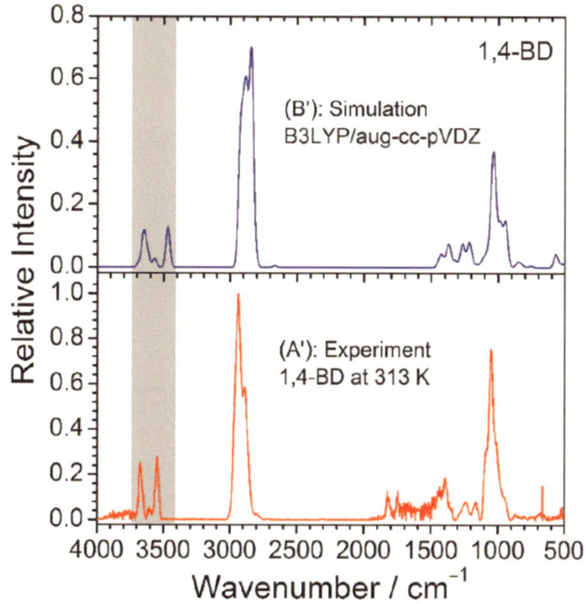

Figure 11.
Experimental IR absorption spectra of 1,4-BD in the gas phase at 313 K (A′), stimulated spectrum of a mixture of conformers g′GG′Gt, gG′G′Gt, tG′TGt, g′TTGt, gGTGt, tTTTt, tGGGt, tTGG′t, gTGGt, and tTGTt (B′). B3LYP/aug-cc-pVDZ calculated anharmonic intensities in the individual spectra of conformers were weighted by the population of the respective conformer [10]. The simulated spectrum (B′) was obtained using Gaussian functions centered at the anharmonic frequencies and with a band width at half-height of 35 cm^{-1}. The O-H stretching region marked with gray. Reproduced from reference [10] with permission. Copyright (2015) ACS.

B3LYP/aug-cc-pVDZ method were found to be 12.8, 4.4, 21.8, 26.8, 18.8, 3.3, 1.9, 5.1, 3.9, and 1.2% for g′GG′Gt, gG′G′Gt, tG′TGt, g′TTGt, gGTGt, tTTTt, tGGGt, tTGG′t, gTGGt, and tTGTt conformers, respectively. It has been found that most stable hydrogen bonded conformers of 1,4-BD are less populated than some of the non-hydrogen bonded conformers. Even for the 1,4-BD, the relative population of the g′GG′Gt conformer [10], which has strong intramolecular hydrogen bond, is less than what is predicted. It has been proposed that perhaps the intramolecular hydrogen bond is not the only factor governing the relative stability of the hydrogen bonded conformers of diols.

This comparative study has revealed a few notable things. First of all, experimental observed spectra are a mixture of hydrogen and non-hydrogen bonded conformers of diols. Second thing is that it helps in finding what is the weighting factor by which individual conformer contributed to the measure spectra that indirectly indicating in what % each conformer exist in a mixture. Furthermore, based on this comparative study, it has been concluded that strong intramolecular hydrogen bonding exists in 1,4-BD but it appears to be weak intramolecular hydrogen bonding in 1,2-ED at temperatures of 303, 313 and 323 K in the gas phase [10].

4.3 Photochemistry with time resolved set-up

Using gas phase time-resolved FT-IR set-up described in **Section 2.6**, the very first system that has been studied by Behera et al. [2] is the photodegradation of

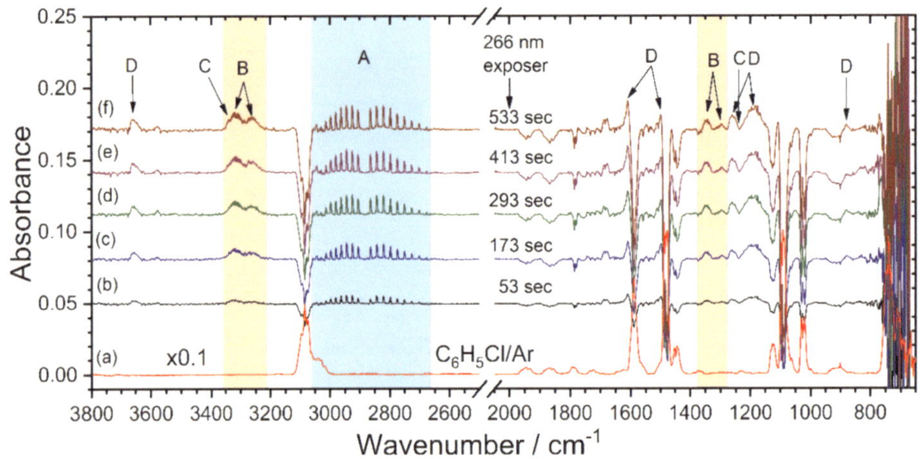

Figure 12.
(a) Infrared absorption spectra of 0.43 ± 0.01 torr C_6H_5Cl seeded in 142 torr argon (Ar). Difference spectra (b–f) were measured during 53, 173, 293, 413, and 533 sec photolysis at 266 nm. Infrared absorption band indicated by A, B, C, and D are assigned to hydrochloric acid (HCl), acetylene (C_2H_2), 1,3-butadiyne (C_4H_2), and phenol (C_6H_5OH), respectively. Reproduced from reference [2] with permission. Copyright (2021) Elsevier.

chlorobenzene (C_6H_5Cl). During this measurement, spectrometer has been set to acquired 5 spectra at 120 sec intervals for a total period of about 9 min with a spectral resolution 2 cm^{-1} and averaged over 128 scans. **Figure 12a** shows the spectra of precursor (C_6H_5Cl) which is obtained based on normal scan measurement, whereas spectra in **Figure 12b–f**, obtained during 53, 173, 293, 413, and 533 s photolysis of C_6H_5Cl were based on repetitive mode measurement.

Following 266 nm photolysis of C_6H_5Cl, the ro-vibrational lines were observed in the region 3060–2625 cm^{-1}, at 3317.8/ 3262.7 cm^{-1} and 1346.2/1301.2 cm^{-1}, and at 3341.2 and 1232.7 cm^{-1}. These infrared features are assigned to the hydrochloric acid, acetylene, and 1,3-butadiyne, respectively. Identification of acetylene and 1,3-butadiyne but not expected HCl co-product *ortho*-benzyne (o-C_6H_4) indicates, possibly, o-C_6H_4 further degraded into acetylene and 1,3-butadiyne. The calculated potential energy surfaces for the possible degradation channels of C_6H_5Cl shown that HCl elimination and C-Cl bond fission are major degradation paths.

Further quantitative analysis has done using the measured infrared absorbance for the precursor and various photoproducts and Eq. (5). The results of this analysis have been presented in **Figure 13**.

Accordingly, the initial pressure of precursor is estimated to be 0.427 ± 0.013 torr. During first 53 s photolysis the concentration of C_6H_5Cl is decreased by $(0.0128 \pm 0.241 \times 10^{-3})$ torr and formation of primary product HCl is estimated to be $(0.00645 \pm 0.19894 \times 10^{-4})$ torr as reflected in **Figure 13**. This clearly indicates that, the experimental branching ratio for the HCl elimination and C-Cl bond fission paths of C_6H_5Cl is 1:1. This is consistent with the calculate energetics which indicates that the energy required for the C-Cl bond fission is closed to the HCl elimination barrier of C_6H_5Cl. Further details assignment of photodegraded products and reaction mechanism can be found in recent publication [2].

This gas phase time resolved technique was successfully applied to investigate the photodegradation of C_6H_5Br at 266 nm.

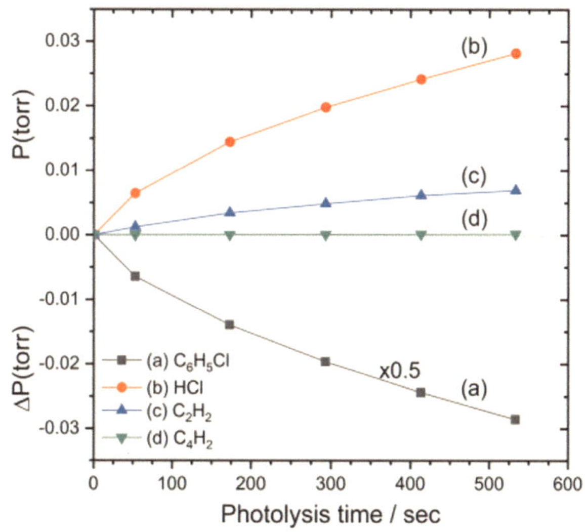

Figure 13.
(a) Decay of chlorobenzene (C_6H_5Cl) and growth of products (b) hydrochloric acid (HCl), (c) acetylene (C_2H_2), and (d) 1,3-butadiyne (C_4H_2) as a function of 266 nm photolysis time. The plot is based on repetitive mode measurements, see ***Figure 12*** *and Eq. (5) in the main text. Reproduced from reference [2] with permission. Copyright (2021) Elsevier.*

5. Conclusions

These studies have provided an experimental verification of low concentration infrared spectroscopic measurements using multi-pass long-path gas cell coupled with the FT-IR spectrometer. This is essential for the detection of atmospherically and astronomically important PAHs and MPAHs. The low concentration measurement is also essential for the detection of intramolecular hydrogen bonded conformers in a gaseous mixture. Furthermore, even its required for the detection of photoproducts in a mixture of reactant and products.

The benchmark data on vibrational signatures of some of the MPAHs are expected to help in identification of them in a complex mixture from various sources. These studies also provide the spectroscopic method of quantification of gaseous MPAHs which will eventually help in quantification of them so that worldwide emission rate and adverse effects on human being can be monitored.

The gas phase infrared spectra are expected to provide more information as compared to solid or liquid state spectra. This has been inferred based on comparison of gas phase spectra and its corresponding liquid phase spectra of diols. Also, experimental spectra can be reconstructed after properly weightage of calculated spectrum for various conformers which helps in finding % contribution of various conformers of diols in a complex mixture. Perhaps, this is something, similar to the method called multivariate analysis.

The repetitive mode time resolved FT-IR coupled with laser was successful to probe the HCl elimination channel in the UV photolysis of C_6H_5Cl combined with *ab* initio calculations. The secondary products C_2H_2 and C_4H_2 was observed in conjunction with the primary product HCl. These secondary products play important roles to obscure the results from the primary processes particularly when low time resolution measurement was involved. The experimental conditions like continuous photolysis of chlorobenzene seeded in a buffer gas mimic the reaction in the atmosphere.

The future potential application such technique is in the pharmaceutical industries to find the drug degradation. The study of the photodegradation behavior of pharmaceuticals in the environment is a key issue in terms of the formation of toxic products. Therefore, information on drug degraded products is crucial to understand the environmental fate of contaminants and to establish the important degradation pathways. This can be done using this low time-resolved FT-IR spectroscopy.

Acknowledgements

My sincere thanks to Prof. Puspendu K. Das and Prof. E. Arunan of the Department of Inorganic and Physical Chemistry (IPC), Indian Institute of Science (IISc), Bangalore, for allowing us to use their lab and departmental facilities. My sincere thanks to Dr. Shubhadip Chakraborty and Dr. Bedabyas Behera for their contribution of some part of their work as cited. The spectrometer used in the experiment was supported by the FIST program of the Department of Science and Technology, Govt. of India. Dr. Das thank Center for Advanced Research Studies, Ganpat University (CARS-GUNI) for providing working platform. Also, thanks to all authors for given permission to use their published work either in the form of data or figure.

Conflict of interest

The authors declare no conflict of interest.

Author details

Prasanta Das
Ganpat University, Mehsana, Gujarat, India

*Address all correspondence to: pdas2004@gmail.com

References

[1] Hughey KD, Tonkyn RG, Harper WW, Young VL, Myers TL, Johnson TJ. Preliminary studies of UV photolysis of gas-phase CH3I in air: Time-resolved infrared identification of methanol and formaldehyde products. Chemical Physics Letters. 2021;**768**:138403. DOI: 10.1016/j.cplett.021.138403

[2] Behera B, Das P. HCl elimination in the photolysis of chlorobenzene at 266 nm: An FT-IR spectroscopy and Quantum Chemical Study. Chemical Physics Letters. 2021;**774**:138601. DOI: 10.1016/j.cplett.2021.138601

[3] Mezzetti A, Schnee J, Lapini A, Donato MD. Time-resolved infrared absorption spectroscopy applied to photoinduced reactions: How and why. Photochemical & Photobiological Sciences. 2022;**21**:557-584. DOI: 10.1007/s43630-022-00180-9

[4] Mezzetti A, Leibl W. Time-resolved infrared spectroscopy in the study of photosynthetic systems. Photosynthesis Research. 2016;**131**:121-144. DOI: 10.1007/s11120-016-0305-3

[5] Huang YH, Chen JD, Hsu KH, Chu LK, Lee YP. Transient infrared absorption spectra of reaction intermediates detected with a step-scan fourier-transform infrared spectrometer. Journal of The Chinese Chemical Society. 2014;**61**:47 58. DOI: 10.1002/jccs.201300415

[6] Smith GD, Palmer RA. Fast time resolved mid-infrared spectroscopy using an interferometer. In: Chalmers JM, Griffiths PR, editors. Handbook of Vibrational Spectroscopy. Theory and Instrumentation. Chichester: Wiley; 2002. pp. 625-640

[7] Rakib F, Al-Saad K, Ahmed T, Ullah E, Barreto GE, Ashraf GM, et al. Biomolecular alterations in acute traumatic brain injury (TBI) using Fourier Transform Infrared (FT-IR) imaging spectroscopy. Spectrochimica Acta Part A: Molecular and Biomolecular Spectroscopy. 2021;**248**:119189. DOI: 10.1016/j.saa.2020.119189

[8] Bec KB, Grabska J, Huck CW. Biomolecular and bioanalytical application infrared spectroscopy - A Review. Analytica Chimica Acta. 2020; **1133**:150-177

[9] Wrobel T, Bhargava R. Infrared spectroscopic imaging advances as an analytical technology for biomedical sciences. Analytical Chemistry. 2017;**90**: 1444-1463. DOI: 10.1021/acs.analchem.7b05330

[10] Das P, Das PK, Arunan E. Conformational stability and intramolecular hydrogen bonding in 1,2-Ethanediol and 1,4-Butanediol. The Journal of Physical Chemistry. A. 2015; **119**:3710-3720. DOI: 10.1021/jp512686s

[11] Sonseca A, Fray M El, RSC Advances 2017;7:21258-21267. DOI: 10.1039/c7ra02509b

[12] El-Azazy M. Introductory Chapter: Infrared spectroscopy – A synopsis of the fundamentals and applications. In: El-Azazy M, editor. Infrared Spectroscopy: Principles, Advances, and Applications. London: IntechOpen; 2018. pp. 1-7. DOI: 10.5772/intechopen.82210

[13] Griffiths PR, Haseth JA. Fourier Transform Infrared Spectroscopy. 2nd ed. New Jersey: Wiley & Sons, Inc; 2007

[14] Michelson AA. Visibility of Interference-fringes in the focus of a telescope. The Philosophical Magazine. 1891;**31**:256-259. DOI: 10.1080/14786449108620101

[15] Das P. Gas Phase Infrared Spectra of PAHs and Diols: Experiment and Theory. Bangalore: Indian Institute of Science; 2011

[16] Herriott DR, Schulte HJ. Folded optical delay lines. Applied Optics. 1965; **4**:883-889. DOI: 10.1364/AO.4.000883

[17] Das P, Chakraborty S. Gas Phase Infrared Spectra of PAHs and Methylated-PAHs. Germany: LAP LAMBERT Academic Publishing; 2018

[18] Behera B, Das PK. Blue- and red-shifting hydrogen bonding: A gas phase FT-IR and Ab initio study of RR′CO ... $DCCl_3$ and RR′S ... $DCCl_3$ complexes. The Journal of Physical Chemistry. A. 2019;**123**:1830-1839. DOI: 10.1021/acs.jpca.7b11962

[19] Chakraborty S, Das P, Manogaran S, Das PK. Vibrational spectra of fluorine, 1-methylfluorene and 1,8-dimethylfluorene. Vibrational Spectroscopy. 2013;**68**:162-169. DOI: 10.1016/j.vibspec.2013.07.001

[20] Das P, Arunan E, Das PK. Infrared spectra of dimethylphenanthrenes in the gas phase. Journal of Physical ChemistryA. 2012;**116**:5769-5778

[21] Behera B, Das P. A spectroscopic method for monitoring photochemical reactions in the gas phase. MethodsX 2022

[22] Eberhard J, Yeh PS, Lee YP. Laser-photolysis/time-resolved Fourier-transform absorption spectroscopy: Formation and quenching of HCl(ν) in the chain reaction $Cl/Cl_2/H_2$. The Journal of Chemical Physics. 1997;**107**: 6499-6502. DOI: 10.1063/1.474264

[23] Stoppa P, Charmet AP, Tasinato N, Giorgianni S, Gambi A. Infrared spectra, integrated band intensities, and anharmonic force field of $H_2C{=}CHF$. The Journal of Chemical Physics. 2009;**113**: 1497-1504

[24] Galabov BS, Dudev T. Vibrational intensities. In: Durig JR, editor. Vibrational Spectra and Structure. Amsterdam: Elsevier; 2002. pp. 1-23

[25] Das P, Manogaran S, Arunan E, Das PK. Infrared spectra of dimethylquinolines in the Gas Phase: Experiment and Theory. The Journal of Physical Chemistry. A. 2010;**114**: 8351-8358. DOI: 10.1021/jp1056896

[26] Robertson WW, Matsen FA. Molecular orbital theory and the near ultraviolet adsorption spectrum of monosubstitute benzenes. IV. The phenyl halides and the inductive effect. Journal of American Chemical Society. 1950;**72**:5252-5256. DOI: 10.1021/ja01167a130

[27] Chakraborty S, Das P, Das PK. Isomeric identification of methylated naphthaenes using gas phase infrared spectroscopy. Indian Journal of Physics. 2012;**86**:209-218. DOI: 10.1007/s12648-012-0042-1

[28] Das P, Arunan E, Das PK. Infrared spectra of dimethylnaphthalenes in the gas phase. Vibrational Spectroscopy. 2008;**47**:1-9. DOI: 10.1016/j.vibspec.2008.01.002

[29] Behera B, Chakraborty S. Room Temperature gas phase infrared spectra of H-bonded oligomers of methanol. Vibrational Spectroscopy. 2020;**106**: 102981. DOI: 10.1016/j.vibspec.2019.102981

[30] Jesus AJL, Rosado MS, Reva I, Fausto R, Eusebio MES, Redinha JS. Structure of isolated 1,4-butanediol: Combination of MP2 calculations, NBO analysis, and matrix-isolation infrared spectroscopy. The Journal of Physical

Chemistry. A. 2008;**112**:4669-4678. DOI: 10.1021/jp7116196

[31] Howard DL, Jorgensen P, Kjaergaard HG. Weak intramolecular interaction in ethylene glycol identified by vapor phase OH-stretching overtone spectroscopy. Journal of the American Chemical Society. 2005;**127**:17096-17103. DOI: 10.1021/ja055827d

[32] Cramer CJ, Truhlar DG. Quantum chemical conformational analysis of 1,2-ethanediol: Correlation and solvation effects on the tendency to form internal hydrogen bonds in the gas phase and in aquesous solution. Journal of the American Chemical Society. 1994;**116**: 3892-3900. DOI: 10.1021/ja00088a027

Chapter 4

IR Spectroscopy in Qualitative and Quantitative Analysis

Nabeel Othman

Abstract

The infrared technique is one of the oldest techniques; it deals with the frequencies of bond vibration in a molecule. The main uses of this technique are to identify and determine components in various organic or inorganic compounds. In this technique, a part of the incident infrared radiation is absorbed by the molecules of the sample and the other is transmitted. The favorite method of infrared spectroscopy is FTIR (Fourier transform infrared). There have been many developments in using IR technique in qualitative and quantitative analyses, including the first and second derivatives of the infrared spectrum. IR rays do not damage the exposed skin like other rays such as ultraviolet light. It must be mentioned that the IR technique was used in hyphenated techniques (instead of the detector in chromatographic device), for example, after separation by gas chromatography detected by IR. Also, this chapter contains essential information about Raman spectroscopy. Infrared spectroscopy is a technique that has acceptable accuracy and sensitivity to be one of the most important analytical techniques used in the qualitative analysis, and also, it is used in the quantitative estimation of compounds through measuring the transmitted or absorption intensity of the active groups.

Keywords: infrared, Raman, first and second derivatives, qualitative and quantitative estimations, hyphenated techniques

1. Introduction

The introduction included the followings below:

1.1 Infrared spectroscopy

Spectroscopy is the branch of science contracts with learning about the interaction of the radiation of electromagnetic rays with substances.

Electromagnetic Radiation (EMR) is a type of energy that is around us and taking various forms, these types included radio waves, microwaves, infrared, visible light, ultraviolet X-rays, and gamma-rays. Sunlight is also considered a form of EMR, with Vis light only a minor share of the EM spectrum, which covers a wide range of wavelengths. Visible light has high energy compared with IR light [1, 2].

Infrared Spectroscopy (IRS) deals with the frequencies of bond vibration in a molecule. The main use is to identify the functional groups in many samples. The most covalently bonded compounds, whether organic or inorganic compounds, absorb electromagnetic radiation in the region of infrared. This IR region lies between the visible light and the microwaves region. IR radiation mainly considers thermal energy, in covalent bonds it gives stronger vibrations to molecules. Near-IR can be used in direct determination (nondestructively) of protein present in feeds, and this type of IR region is increasingly used in analytical chemistry for quantitative analysis of various compounds [1].

IR can be divided into main three different bands:

1. Near-Infrared (NIR, 0.78~3.0 μm).

2. Mid-Infrared (MIR, 3.0~50.0 μm)

3. Far-Infrared (FIR, 50.0~1000.0 μm) [3].

In UV and Vis. of the spectrum, the unite of wavelength is nanometer (nm), while in the infrared region wavenumbers are used, and cm^{-1} is the unit [2, 4]. The IR spectrum is drawn via a plot of absorbed or transmittance% (T%) against the wave-number (**Figures 1** and **2**).

1.2 Fourier transform infrared spectroscopy (FTIR)

The favorite method of IRS is FTIR (Fourier Transform infrared), in IRS the infrared radiation is passed through the investigated sample. A part of the incident infrared radiation is absorbed by the sample and the other is transmitted. The resulting spectrum represents the absorption molecules. FTIR spectrophotometers have many advantages when compared with the older techniques IR, the FTIR instruments are more accurate, and more sensitive, all frequencies of functional groups are estimated simultaneously compared with an individual estimation of functional groups in IR, and they are fast in performance as was in the case of older IR instruments.

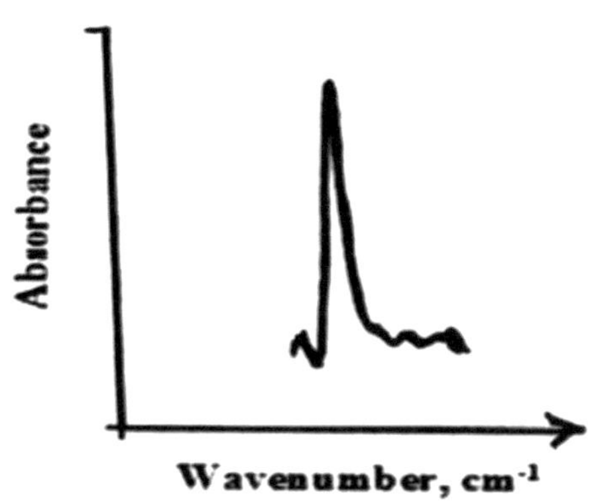

Figure 1.
The spectrum of an absorption mode.

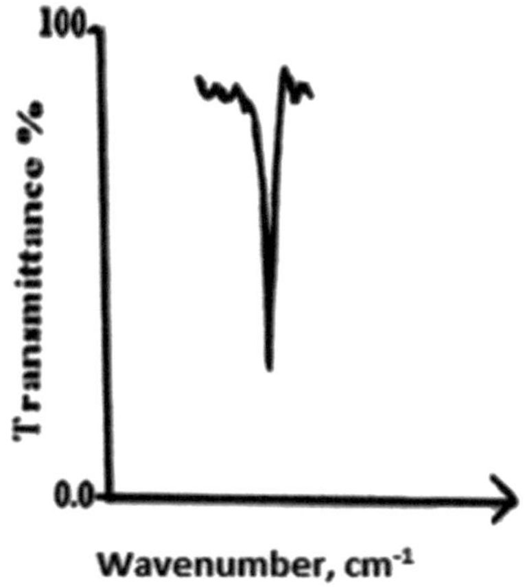

Figure 2.
The spectrum of T% mode.

1.3 Classification of IR bands

Figure 3 shows the main three types of IR bands they classified according to their relative intensities in the IR spectrum.

An increase in the dipole moment according to the increase in the distance between atoms caused an increase in the intensity of the absorption peak [5].

1.4 IR peaks shapes

Two main types of IR band shapes are narrow (thin and pointed) and broad (wide and smoother). An example for broad is the O-H peak in alcohols and carboxylic acids, as shown below in **Figure 4** [5].

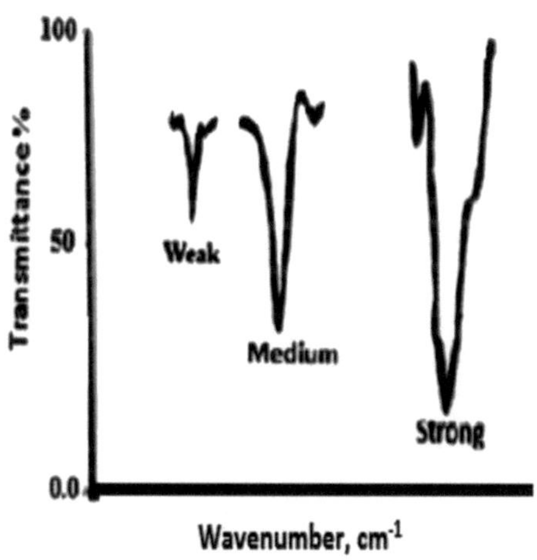

Figure 3.
The types of IR bands according to their relative intensities.

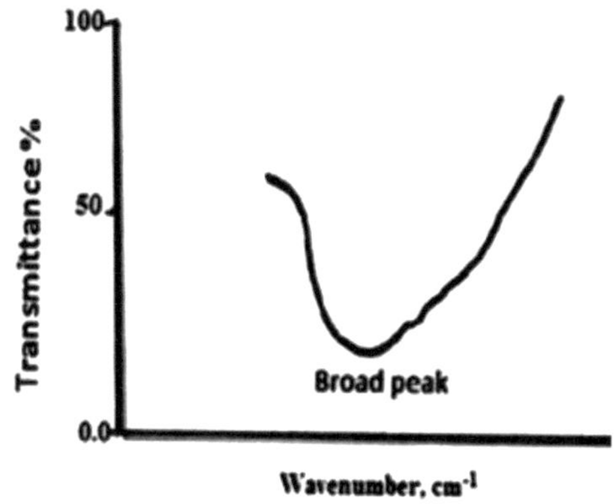

Figure 4.
The broad peak of the hydroxyl group.

1.5 Range of IR absorption

The typical IR absorption range to covalent bonds in molecules is from 600 to 4000 cm^{-1}. The graph shows the regions of the spectrum where the following types of bonds normally absorb. For example, the sharp band around 2200–2400 cm^{-1} would designate the possibility of the presence of a C-N or a C-C triple bond, and other ranges in IR-absorption for other types of bounds.

1.6 Overtones and combination bands

When a molecule absorbed electromagnetic radiation in the IR region, then the molecule is promoted from the ground state to the second, third, or even fourth vibrational excited state. These bands are known as Overtones. The intensity of these bands is very weak. It is helpful in the characterization of aromatic compounds.

When two fundamental vibrational frequencies ($\nu 1 + \nu 2$) in a molecule couple give rise to a new vibrational frequency within the molecule, it is known as a combination band.

1.7 Coupled vibrations

The coupled vibrations are observed in groups such as $-CH_2$, NH_2, etc. In these groups, the same atoms are attached to the central atom. When $-CH_2$ undergoes vibration, vibrational frequencies for the $-CH_2$ group are observed at 2950 cm^{-1} (asymmetric stretching) and 2860 cm^{-1} (symmetric stretching). A number of molecules contain the same functional group and show a similar peak above 1500 cm^{-1}, but they show a different peak in the fingerprint region. Therefore, we can say that each and every molecule has a unique peak or band, which is observed in the fingerprint region; it is just like the fingerprint of a human.

1.8 The functional groups and fingerprint regions

IR spectrum can be separated mainly into two regions. Most of the functional groups show absorption bands at the wavelength (4000–1200 cm^{-1}) region, which is called the functional group region. Will the second region from 1200 to 400 cm^{-1} is called the fingerprint region. Fingerprint region is characteristic of the compound as a whole. An example is 2-pentanol and 3-pentanol, the two compounds with similar absorption in the functional group region. However, their fingerprint regions are different, because the two compounds differ, and to accurately identify the compound by comparing the fingerprint area with the fingerprint area of a standard or known sample of this compound [6].

1.9 Factors affecting the vibrational frequency

The main factors affecting the vibrational frequency are listed below:

A. Conjugation: As the conjugation increases, stretching frequency decreases, because force content decreases due to conjugation.

B. Inductive effect and resonance effect: Oxygen is more electronegative than nitrogen; therefore, nitrogen easily donates electron or ion pair of nitrogen undergoes delocalization with a C=O bond. Due to delocalization double bond of a C=O change into a partial double bond, therefore force constant decreases, which decreases the C=O stretching frequency.

C. Hydrogen bonding: Intermolecular hydrogen bonding weakens the O-H bond, thereby shifting the band to a lower frequency. For example, in a clear solution O-H stretching vibration of phenol was observed in the range from 3400 to 3300 cm^{-1}. When the solution is diluted the O-H frequency shifted toward a higher frequency at 3600 cm^{-1}. Whereas in the case of methyl salicylate, intramolecular hydrogen bonding lowers the stretching frequency of O-H at 3200 cm^{-1}. Intramolecular hydrogen bonding does not change its frequency even in a very dilute solution because upon dilution structure of the compound does not change.

D. Ring strain: As the size of the ring decreases, the vibrational frequency of C=O increases. For example [5]:

Cyclohexanone	Cyclopentanone	Cyclobutanone
1710 cm^{-1}	1745 cm^{-1}	1780 cm^{-1}

An increase in wavenumber of the carbonyl group

1.10 General uses of IR

- One of the most important uses of infrared rays is for military purposes, and one of these uses is in binoculars for night vision in case of difficulty in seeing and observing hostile targets.

- Use in remote sensing, astronomy, and space in planetary detection, radio communications, spectroscopy, and weather forecasting.

- Infrared radiation, which is the oldest technique used in wireless communication, and is used in remote control and TV or recorder, as it is used in calculators, one of disadvantages is the speed offered is slow compared with other wireless technologies.

- Spectroscopy Infrared is a widely used technique to help identify carbon-containing organic compounds. Only the polar molecules are active because they have a permanent dipole moment.

A molecule to absorb IR, the vibrations or rotations within a molecule must cause a net change in the dipole moment of the molecule. The principle of action is to shine infrared light so that it passes through the organic compound to be identified; absorption occurs for some of the frequencies by the model. The different precise frequencies of absorption can be used to identify the different groups in the unknown compound, which represent specific groups of atoms within the molecules over a period of time. We can identify the compound because each group has an absorption frequency that differs from the other. Using a detector to determine the different absorbance, which records the amount of infrared light that passes through the compound. Some frequencies pass without being fully absorbed, while others will be greatly absorbed due to the special chemical bonds in the molecules. This leads to obtaining a spectrum containing different selves expressing the totals in the model [7].

- Infrared therapy numerous studies have been described that IR can recover the healing of skin wounds, relieve pain, psychiatric disorders, and cardiac stem cells. There are two types of treatments:

  1. Low-level light therapy (LLLT) using light of low power intensity and the effects are not a response to heat but to the light. The popular light sources used are low-power lasers.

  2. Photobiomodulation (PBM) therapy uses non-ionizing types of light sources, including lasers, it is a non-thermal process.

It is now approved that the PBM therapy is an extra accurate and exact term for the therapeutic application of low-level light compared with "LLLT." A basic principle called the biphasic dose-response included that the large doses of light were found to be less actual than smaller doses. The human skin is reliably exposed to environmental IR radiation, which indirectly or directly stimulates the manufacture of free radicals or reactive oxygen species(ROS). 8~12 μm IR radiation is almost used on full-thickness skin wound therapeutic in rats.

IR light crosses the outer layers of the skin and reaches the tissues of the body. The good thing about using infrared light in therapy is that IR rays do not damage the exposed skin like other rays such as ultraviolet light. An advantage of exposure to IR ray that it improves the circulation of blood and promotes cell regeneration [8–11].

1.11 Raman spectroscopy

Raman scattering firstly was observed by Raman and Krishnan (Indian physicists) in 1928. It is an analytical technique where the scattered light is used to measure the vibrational energy styles of molecules. Raman spectroscopy can offer chemical structural information, as well as identify the substances to be studied through their characteristic Raman "fingerprint." Raman spectroscopy extracts the information over the detection of Raman scattering from the investigated sample. After the light is scattered via molecule, the oscillating electromagnetic field of the photon persuades a polarization of the molecular electrons cloud. The photon is transported to the molecule, due to the formation of a very short-lived complex (photon-molecule), and it is called commonly the virtual state. It is not stable and the photon can be re-emitted immediately as scattered light. Approximately 1/10 million photons Raman scattering occurs. The transfer of energy between the scattered photon and molecule and if the molecules gain energy from the photon according to the scattering (an excitation to a higher vibration level) and after that, the scattered photon loses energy, and this phenomena is called Stokes Raman, included an increase in wavelength. If the molecule loses energy by transferring to a lower vibrational level the scattered photon gains energy, inversely, the wavelength decreases, which is called Anti-Stokes Raman. Finally, if most of the molecules are in the ground vibrational level (Boltzmann distribution) and as a result, the Stokes Raman scatter is a continuously more probable process and intense than the anti-Stokes; for this reason, it is approximately always the Stokes Raman scatter used in Raman spectroscopy.

The main differences between IR and Raman scattering are listed in **Table 1**.

As a common rule included that everything that does not seem in the IRS is taken in Raman (bond of molecule either be with Raman active or be IR active but it not with be both). H_2 or CCL_4 doesn't have spectrum in IR; but they give spectra in Raman. Also nitrogen-nitrogen, carbon-carbon, and sulfur-sulfur bonds have a change in polarizability, the incident photons interact with these models, these are examples of bonds that give rise to Raman active spectrum bands, but it is difficult to get spectrum in FTIR [12, 13].

Raman spectroscopy has several applications, such as the identification of materials and identification of different minerals ranging from iron oxy (hydroxides) to rare minerals. Study of the crystallinity, the composition, and uniformity, and also measurement of local temperature and stress. Raman spectroscopy is nondestructive, and the technique has a good resolution [14].

Recently, Raman spectroscopy has been used in blood identification and distinguishing between human and nonhuman blood using a portable Raman spectrometer, which can be used at a crime division, and the bloodstain of human could be

No.	IR	Raman
1.	The principle based on the light absorption.	The principle based on scattering of light.
2.	To appear the spectrum the variation in the polar moment of the molecule to be study must not be equal to zero.	To achieve the Raman spectrum it is not important to have dipole moment or the change of polarity, the bonds of molecular have specific transition energy in which cause a change of polarizability to give a rise to Raman active.
3.	The source of light used depend on the region of electromagnetic spectrum, tungsten filament lamp in Near-infrared, coil of Nichrom wire in Mid- infrared and high pressure mercury-arc lamp in Far infrared.	laser was the excitation source. Almost , solid state lasers types are used in Raman tools with general wavelengths of 532, 785, 830 and 1064 nm.

Table 1.
The main differences between IR and Raman spectroscopy.

distinguished from the non-human ones via using a principal component analysis, and also this analysis is useful for forensic [15].

2. Application

2.1 Qualitative analysis

Example 1:

FTIR spectroscopy is the most reliable tool for identifying bone types and can also be widely used in forensic medicine. Identification of human and non-human skeletal remains unknown to investigators and is of great interest in forensic and anthropological procedures. Especially when the traditional morphological methods for diagnosing and differentiating between these types of bones took a long time. Therefore, the use of infrared spectroscopy and chemical measurement methods to determine the spectral differences between these two types of bones, human, and non-human bones (such as pigs, goats, and cows). The results showed that pig bone is not suspicious of human bone in the study of changes after death because it is more sensitive to environmental conditions than human bone [16].

Example 2:

The micro-FTIR technique was used to characterize the components of a dye painted on the walls of a church in Cyprus. The product was copper-based and the dye contained hydrated copper oxalate. Reflective imaging of the localization sites for the presence of copper and calcium oxalate within the layers of the plate. We conclude from this study that imaging calcium oxalate within different layers of paint samples is very important for studying copper-based pigments in general, and in particular for analyzing pigments used in coatings on different external surfaces [17].

Example 3:

Different heterocyclic compound derivatives have been synthesized via the reaction of ortho-Carboxybenzaldehyde with various aromatic amines (using six amino compounds) to produce Schiff bases (**Figure 5**).

The Schiff bases compounds gave FTIR spectra with an absorption appeared at wavenumber between 1602 and 1614 cm^{-1} this peak belongs to the new C=N group, and also carbonyl of carboxyl group gave absorption appeared at (1741–1766) cm^{-1}, and the absorption at (3306–3462) cm^{-1} for OH group of carboxylic acid. The authors noticed that the carbonyl of aldehyde disappeared, therefore our conclusion that FTIR proves the suggested mechanism and helps to suggest the structure of the product using the absorption of selective functional groups (**Table 2**, **Figure 6**) [18].

2.2 Quantitative analysis

Example 1:

Fourier transform infrared (FTIR) is used in numerous areas of industrial pharmacy with satisfactory results. The technique's characteristic and nature tolerate unequivocally bright forecasts for quantitative analysis. FTIR is considered a green analytical chemistry technique. It is very easy, fast to work by a temperately knowledgeable technician, covers a large range of spectra to analyze the pharmaceutical formulations, the main advantages are that it has a good resolution and is considered nondestructive

DOI: http://dx.doi.org/10.5772/intechopen.106625

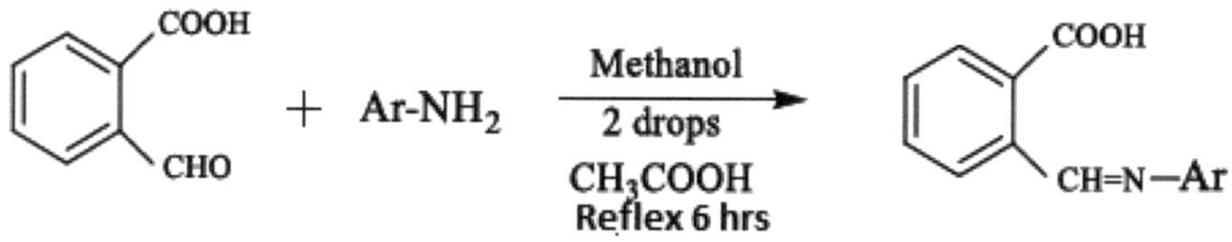

Figure 5.
The reaction of Schiff base preparation.

Compound No.	1	2	3	4	5	6
Ar Structure	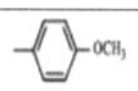	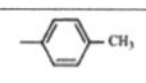	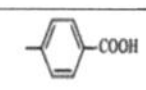	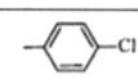	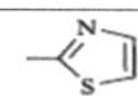	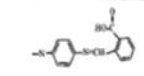

Table 2.
The aromatic part of amine.

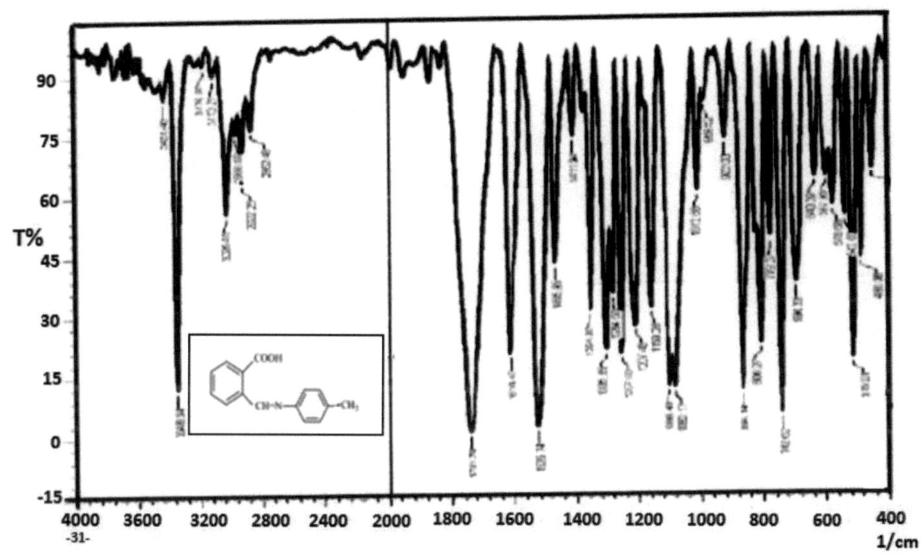

Figure 6.
FTIR spectrum of the product resulted from the reaction of p- toluidine with ortho-carboxybenzaldehyde [from reference 18].

device, and it is also friendly to the environment because in procedure no use of a dangerous organic solvent or any harmful reagents is required for the analysis. Many attempts were suggested for using derivative IR in determination diclofenac sodium in its formulations, but the results indicated that the first derivative spectra are the best technique for determination of diclofenac sodium. The first derivative spectra deleted IR band overlapping with the band understudies and increased sensitivity without any interference of the other band's [19].

Example 2:

Abdulhameed and Nabil (2022) developed a simple and rapid method for the determination of ketoprofen. The method is based on normal and infrared derivative (first derivative) spectroscopy. The results of the study found that the method is accurate and there is the possibility of its application in quality control to determine ketoprofen in pharmaceutical formulations. Ketoprofen was quantified in a range of estimation from 1000 to 4000 μg/ml. This range was based on measuring the T% of the normal spectrum and its first derivative spectrum versus the concentration of ketoprofen in the solution (**Figure 7**). The results prove the validity of the method, as the relative errors were +4.33% and 4.78% and the RSD% values were 1.15% and 1.37%, respectively, and since the values are less than ±5%, the method is considered accurate and precise. The research also included the application of the two methods to estimate the compound under study in its different pharmaceutical preparations with a comparison of the results obtained with the results obtained via using high-performance liquid chromatography technique and calculating the t-student and F tests at P = 0.05.

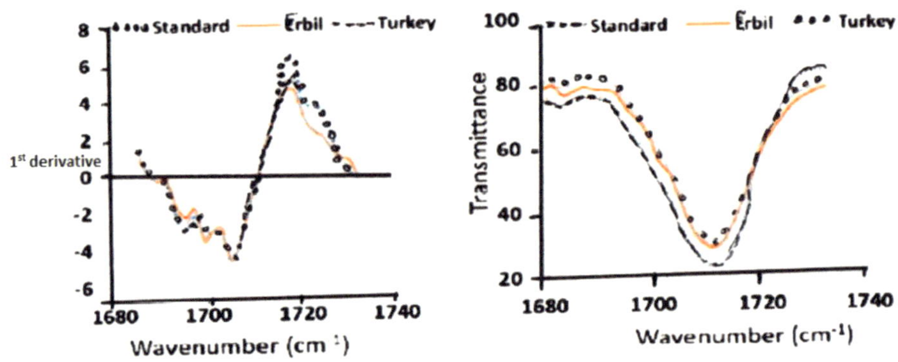

Figure 7.
The first derivative and the normal spectra of two standard ketoprofen from two companies Erbil and Turkey [from reference 20].

Figure 7 shows the derivative spectra of Standard, Erbil, and Turkey ketoprofen solutions, CCl4 was the solvent used. The two individual peaks of carbonyl groups at 1718 cm^{-1} as a positive peak and at 1705 cm^{-1} as a negative peak, and these peaks gave two calibration curves as various concentrations analyses, there is a reverse proportional relationship between the concentration and the percentage of transmittance(T%) (**Figure 8**) and the other indirect proportion. The reverse proportional relationship is according to decreases in the transmittance% of the solution with an increase in concentration (as shown in **Figure 8**), will in **Figure 9** there is a direct proportion or positive relationship for the first derivative IR according to the peak chosen (peaks of carbonyl groups at 1718 cm^{-1} as a positive peak) [20].

Example 3:

Michael et al (1995) used second derivative IR spectroscopy as a non-destructive tool to assess the purity and structural integrity of various samples such as proteins. Spectroscopy using second derivative infrared is a fast, easy, reproducible, cost-effective, and nondestructive method for assessing the purity of samples of some proteins (water-soluble) extracted from a diversity of sources. The 2ed IR spectra were calm under the lab-proven conditions of aqueous (D_2O) solutions of seven different commercial samples for the same enzyme, porcine pancreatic elastase (2.0–3.8 mg protein/100 μl D_2O, pD = 5.4–9.1)., the amide at the region defined by I (1700–1620 cm^{-1}) from the IR spectra using the 2ed derivative for each of the seven elastase

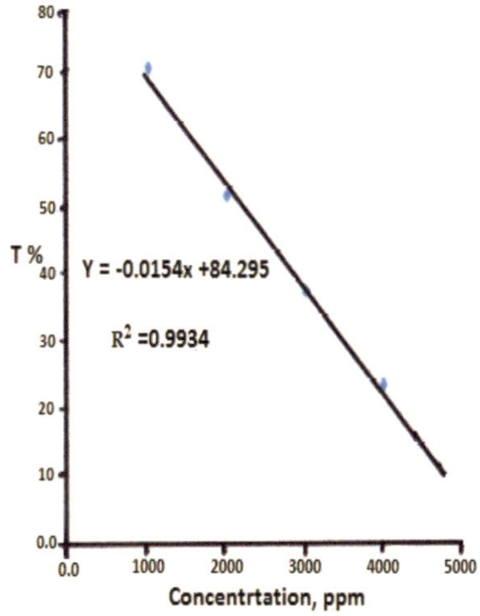

Figure 8.
Calibration curve via normal IR method via first derivative IR.

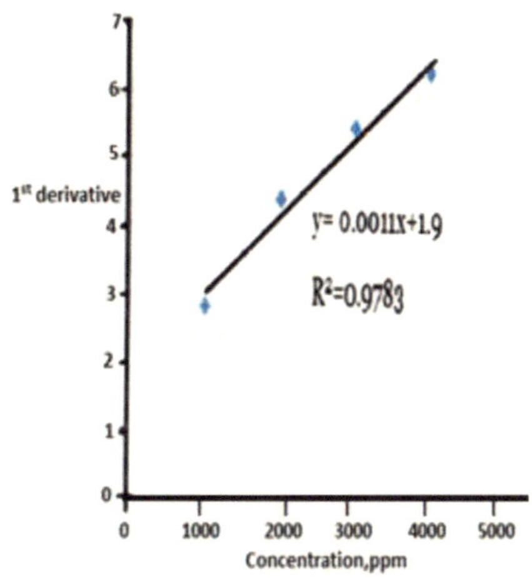

Figure 9.
Calibration curve [from reference 20]. method [from reference 20].

samples displays a characteristic pair of bands: one of them is very weak showing intensities near to 1684 cm^{-1}; the other is close to 1633 cm^{-1} is moderate to strong. While one of the 7 samples under study shows a striking decrease in the noted density of amide I bands relative to the 1516 cm absorbance, along with the appearance of a new strong band at 1614 cm^{-1}. That the seventh sample is of much lower quality than the other samples and sure contains a quintile of the protein present in the non-native state. In addition, the apparent slight changes in the relative location, and intensity of a section of the separate amide I band among the seven spectra indicate slight differences in the formation of the amount of the peptide support of the samples under study. From the results of two samples, it seems that these few changes, sample purity, and identification of non-protein contaminants [21].

3. Hyphenated techniques

During the past five decades, hyphenated techniques developed rapidly and seemed to dominate many analyzes by introducing them to solve many problems related to complex analyzes, as they were widely used in the pharmaceutical industries from the stage of discovery to human use and the study of its concentration in living body fluids. Accuracy and high sensitivity, and one of the most important disadvantages are the high costs of the devices, and they need maintenance and accurate knowledge while working on the device. Liquid chromatography-mass spectroscopy (HPLC -MS) is one of the most widely applied hyphenated techniques because MS is more compatible with high performance-liquid chromatography (HPLC), and has good sensitivity compared with nuclear magmatic resonance (NMR) or IR. It is also possible to connect infrared spectrometers with thermal analyzers, the methods used by thermal analysis give information about the important temperature to study the physical properties of different materials. However, it is not always possible to obtain information about the chemical changes that occur as a result of changes in temperature through the literature. We note that it is possible to link the thermal analyzer with an infrared spectrometer in order to obtain information about the chemical and physical changes that occur at different and more appropriate temperatures. More suitable is the connection between thermogravimetric analysis (TGA) and FTIR spectroscopy However, there are limitations in its analytical use. The more advantages of the hyphenated technique include sensitivity, accuracy, speed, and applicability [7, 22].

3.1 Gas chromatography–infrared (GC-IR)

3.1.1 Difficulties in the combination of GC-IR

In the development of joining the IR technique with GC, the speed of the IR must be changed to a high speed so that the unknown components can enter at the same speed from the GC column, in this case, there is a loss of efficacy and the results are not complete. The best way to solve the connecting problem is that the condensation of the gas that comes out from the column and the process is not easy, it must collect all gas eluted because it contains the component and the gas is collected in a cooled part that converts the gas into a liquid because the infrared technique deals with the liquid solutions. Reentry GC technique combined with Fourier transform infrared to give faster and more accurate technique.

3.1.2 Application

Salerno, et al (2020) suggest an accurate method for determination of illicit drugs via gas chromatography–Fourier transform infrared spectroscopy. According to the increasing number of synthetic molecules that can be used in the illicit drug market, correspondingly they require strong separation and sophisticated analytical techniques. It can be achieved by spectroscopic measurements, using firstly a gas chromatography (GC) technique as the separation device. Then the GC is coupled with FTIR to give a powerful tool. In the current study, the efficacy of GC-FTIR, in achieving elucidation of the structure of 1-pentyl-3(1-naphthyl) indole, known as JWH-018, a synthetic cannabinoid whose components have been identified as being a component of non-incense "incense blends" have been demonstrated in the current study. Moreover, it was quantified with an estimation range on the nano-gram scale. It was obtained in the range of 20–1000 ng, the detection limit and the quantification limit were evaluated to be 4.3 and 14.3 ng, respectively. Finally, the new technique was applied to quantify the activity in the "ST" sample." A real drug seized by law enforcement officers, consisted of a herbal collection containing four types of industrial cannabis belonging to the JWH class. Correct estimation of this type of compound showed that they are chemically similar to each other. The usefulness of the proposed method of analysis using related techniques. It obtains reliable results for complex mixtures of illegal drugs and is a widely applicable alternative to measurement using mass spectrometry [23].

4. Conclusions

Infrared is an important technique and its main application at the beginning to identify polar organic compounds that have a dipole moment. The infrared device has been developed, and we have obtained Fourier transform infrared (FTIR) technique, which is characterized by high accuracy, high sensitivity, and speed of analysis of the compound as a whole. The uses of the technique in the qualitative analysis are identifying the effective groups and the type of bonds between the different atoms constituting the molecule. The technique is used in the quantitative analysis through measurement of absorption or percentage of transmittance (proportional with concentration). The researchers used the first and second derivatives of the infrared spectrum in quantification research and also linked the infrared device with separation devices (for example, GC) to form a new technique called hyphenated techniques, and used in many studies with high sensitivity and precision compared with using each technique individually.

Acknowledgements

I must appreciate thanks and gratitude to the researchers who preceded us for their efforts to reach the information that enriches the subject of infrared techniques, which is the topic of the chapter.

Author details

Nabeel Othman
Chemistry Department, College of Science, University of Mosul, Mosul, Iraq

*Address all correspondence to: nsn20002004@uomosul.edu.iq

References

[1] Donald L, Gary M, George S, James R. Introduction to Spectroscopy. Brooks/Cole, Gengage Learning. Australia. 2009. 745 p

[2] Ralph L, Christine K, Terence C, David Y, Reynold C. The Systematic Identification of Organic Compounds. USA: John Wiley & Sons. Inc; 1956. p. 435

[3] El-Azazy M. Infrared Spectroscopy - Principles, Advances, and Applications Introductory Chapter: Infrared Spectroscopy - A Synopsis of the Fundamentals and Applications. IntechOpen. 2018. DOI: 10.5772/intechopen.82210

[4] Gary D, Purnendu K, Kevin A. Analytical Chemistry 7th ed. United States of America: Petra Recter, John Wiley & Sons, Inc.; 2014. 850 p

[5] Richard A. Factors affecting infrared group frequencies: Carbonyl stretching absorption bands. Applied Spectroscopy. 1986;**40**(3):336-339

[6] Paula Y. Organic Chemistry. 8th ed. USA: Pearson Education; 2016. p. 1340

[7] Douglas A, Skoog F, James H, Stanley R. Principles of Instrumental Analysis. 7th ed. USA: Cengage Learning; 2016. p. 985

[8] Vatansever F, Hamblin M. Far infrared radiation (FIR): Its biological effects and medical applications. Photonics Lasers Medicine. 2012;**4**:255-266. DOI: 10.1515/plm-2012-0034

[9] Tsai S, Yin R, Huang Y, Sheu B, Lee S, Hamblin M. Low-level light therapy potentiates NPe6-mediated photodynamic therapy in a human osteosarcoma cell line via increased ATP. Photodiagnosis and Photodynamics Therapy. 2015;**12**(1):123-130. DOI: 10.1016/j.pdpdt.2014.10.009

[10] Huang Y, Sharma S, Carroll J, Hamblin M. Biphasic dose response in low level light therapy an update. Dose-Response. 2011;**9**(4):602-618. DOI: 10.2203/dose-response.11-009

[11] Shang T, Michael R. Biological effects and medical applications of infrared radiation. Journal of Photochemical and Photobiology. 2017;**170**:197-207

[12] Willard H, Merritt L, Dean J. Instrumental Method of Analysis. 5th ed. New York: D. Van Nostrand and Company; 1974

[13] De Wolf I, Jian C, Spengen W. The investigation of microsystems using Raman spectroscopy. Optics and Lasers in Engineering. 2001;**36**(2):213-223

[14] Das S, Hendry MJ. Application of Raman spectroscopy to identify iron minerals commonly found in mine wastes. Chemical Geology. 2011;**290**(3-24):101-108

[15] Fujihara J, Fujita Y, Yamamoto T, Nishimoto N, Kimura-Kataoka K, Kurata S, et al. Blood identification and discrimination between human and nonhuman blood using portable Raman spectroscopy. International Journal of legal medicine. 2017;**131**(2):319-322. DOI: 10.1007/s00414-016-1396-2

[16] Wang Q, Li W, Liu R, Zhang K, Zhang H, Fan S, et al. Human and non-human bone identification using FTIR spectroscopy. International Journal of Legal Medicine. 2019;**133**(1):269-276. DOI: 10.1007/s00414-018-1822-8

[17] Nevin A, Melia J, Osticioli I, Gautier G, Colombini M. The identification of copper oxalates in a 16th century Cypriot exterior wall painting using micro FTIR, micro Raman spectroscopy and gas chromatography-mass spectrometry. Journal of Cultural Heritage. 2008;**9**(2):154-161

[18] Maysaa K, Al-Jawady M. Synthesis of some heterocyclic compounds derived from ortho-carboxybenzaldehyde. Egyptian Journal of Chemistry. 2022;**65**(1):143-150

[19] Fahelelbom K, Saleh A, Mansour R, Sayed S. First derivative FTIR spectroscopic method as a green tool for the quantitative determination of diclofenac sodium tablets. F1000Research. 2020;**9**:1-14

[20] Abdulhameed M, Nabil A. Determination of ketoprofen in tablet dosage forms by derivative IR spectroscopy. Egyptian Journal of Chemistry. 2020;**65**:215-219

[21] Michael D, Rachel M, Cynthia S, Theodore D. Second derivative infrared spectroscopy as a non-destructive tool to assess the purity and structural integrity of proteins. Pharmaceutical Research. 1995;**12**:446-450. DOI: 10.1023/A:1016273122944

[22] Hellgeth J. Thermal Analysis–IR Methods, in Handbook of Vibrational Spectroscopy. Chichester, UK: Wiley; 2002. pp. 1699-1714

[23] Salerno T, Donato P, Frison G, Zamengo L, Mondello L. Gas chromatography—fourier transform infrared spectroscopy for unambiguous as a determination of illicit drugs: A proof of concept. Frontiers in Chemistry. 2020;**8**:1-12

Chapter 5

Applications of Near-Infrared Spectroscopy (NIRS) in Fish Value Chain

Sonia Nieto-Ortega, Rebeca Lara, Giuseppe Foti, Ángela Melado-Herreros and Idoia Olabarrieta

Abstract

Near-infrared spectroscopy (NIRS) has undergone a significant evolution in the last years due to the numerous scientific studies that revealed its potential for industrial applications, attracting a growing interest in the food sector. Furthermore, new advances have allowed the reduction in size and cost of the NIR devices, making them appropriate for on-site determinations. The complex structure of the fish value chain, combined to its high market value, makes this sector particularly vulnerable to fraud and adulteration practices. Also, the perishable nature of fish and fish products, as well as the lack of traceability, arises the urgent need for a fast, reliable and portable tool capable of precisely characterizing the quality and authenticity of the product while also ensuring its safety. In this chapter, the capabilities of NIRS combined to several chemometric techniques for quality, authenticity and safety applications are presented through an extensive review of the most recent research works.

Keywords: NIR applications, food testing and analysis, spectral analysis, industrial applications, fish chain, chemometrics

1. Introduction

1.1 Principles of NIR spectroscopy and chemometrics

Near-infrared spectroscopy (NIRS) is a technique that measures the absorption of the electromagnetic region from 750 nm to 2500 nm (**Figure 1a**), between the visible and the mid-infrared (MIR) region. Like MIR spectroscopy and some part of the far-infrared (FIR) spectroscopy, this technique detects molecular vibrations, which gives information about the different chemical bonds within the molecule and its environment [1, 2].

However, there are some differences between these spectroscopic techniques, based on the nature of the vibrational transitions. While in the MIR/FIR region (2500 nm–50,000 nm), almost all fundamental transitions ($v_0 \rightarrow v_1$) can be observed, resulting in sharp and intense spectra, in the near-infrared (NIR) region (750 nm–2500 nm) only the much less intense overtones ($v_0 \rightarrow v_n$) of those transitions appear. NIR absorptions

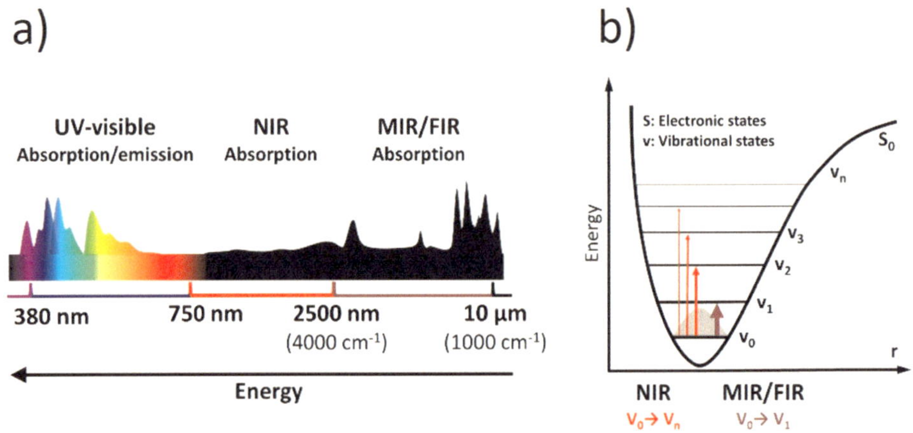

Figure 1.
a) Electromagnetic spectrum of UV-vis-IR. b) Vibrational transitions involved.

are based on overtones and combination bands and, due to their lower transition probabilities, intensities usually decrease by a factor of 10 or 100 from the fundamental overtone to the next one (**Figure 1b**) [1, 3].

It could be logical to think that it would be better to use MIR bands (fundamental transitions) instead of NIR. However, NIR instruments are simpler and more affordable. Furthermore, NIR has a deeper penetration in skin and organic tissues (optical window) than MIR, so that a larger area or volume can be measured [4].

While MIR chemical information can be directly obtained from peak intensities, shape and energy, NIR spectra need special treatment to extract useful information that involves multivariate analysis of data, usually called chemometrics [5]. Chemometrics is the science that applies mathematics and statistical methods to process chemical data, in order to obtain useful information about the sample [6, 7]. These two sciences are closely related: NIRS would not have evolved without chemometrics, and chemometric algorithms usually use NIRS examples to illustrate their power. This process needs computational resources that only since the last few decades have been available. In NIR applications, chemometrics covers a wide range of multivariate methods, which involve preprocessing techniques, as well as qualitative and quantitative analysis [8].

NIRS data usually need to be preprocessed before the model construction, because the spectra usually contain plenty of noise and background information that must be filtered out to find the desired spectral signatures. Some of the most used preprocessing techniques are divided in: i) scatter correction methods and ii) spectral derivatives. The scatter correction methods include techniques such as multiplicative scatter correction (MSC), detrending, normalization, or standard normal variate (SNV). Regarding the spectral derivatives, the Savitzky-Golay (SG) derivative after smoothing is one of the most used [9]. Chemometric methods can be divided into **unsupervised**, when groups or patterns inside data are not known, and **supervised** (classification and regression) when categories or analytes are already quantified, and the objective is to predict them for new samples [10].

The rapid development of chemometric methods during the last decades has allowed the application of NIRS to different areas of food science [11], such as food fraud and authenticity [12, 13] and food quality [14, 15].

DOI: http://dx.doi.org/10.5772/intechopen.105736

1.2 Fish value chain

The demand for fish consumption has increased during the last years as a consequence of the better dietary habits of the population. Fish has become one of the most valuable foods, due to its higher protein content and the healthier fat profile that it has compared with meat [16]. However, fish and fish products present complex problematics in terms of food quality, authenticity and safety. In addition, a greater awareness of food quality and sustainability has led consumers to demand increasingly reliable information about the products purchased in the market, which has stimulated the food value chain to strengthen efforts in the quest for a better and more trusty traceability [17].

Fisheries and aquaculture production are very heterogeneous in terms of species and fish products and, due to its high economic value, are one of the most vulnerable targets to adulteration and fraud. These practices are favoured by the complex structure of the supply chain (**Figure 2**), the lack of transparency and the high percentage of the production destined to the elaboration of processed fish products (around the 50%) such as fillets, portions or elaborated products, in which the morphological characteristics are absent, making these frauds particularly elusive [18]. Furthermore, fish is a perishable product that may suffer a fast enzymatic decomposition and microbial spoilage compared with other foods [19]. This means that small changes in the preservation conditions along the value chain, such as temperature or salt level variations, can have a high impact on the fish quality and safety, resulting in a fast degradation, higher food waste and the risk of intoxications.

On the market side, consumers are demanding more transparency in the labelling of seafood. To ensure the prevention and prompt detection of illegal activities and the access to all the information about the nature, origin and characteristics of the fish, monitoring of all the processes and quality analysis must be done thoroughly through the whole production chain and market, until it arrives to the consumers [20, 21]. The implementation of reliable methods to ensure a complete and correct traceability of the product and to avoid fraud is of primary interest for both industry and consumers.

Figure 2.
Flow chart of seafood value chain from catch to consumer, indicating the different points where the quality of the fish should be monitored. *) Monitoring of chain conditions and quality attributes;* *) inspection technologies and quality control.*

1.3 NIRS in fish analysis

Nowadays, the quality, authenticity and safety control procedures in fish industry are carried out using traditional laboratory techniques, making virtually impossible any real-time application. However, some rapid techniques, such as NIRS, have shown the capacity of obtaining valuable information in a rapid way, which is important for the fish chain analysis [22].

NIRS is a tool already well developed in agriculture [23] and other products [24, 25]. The capacity of this technique to detect vibrations of molecules with polar bonds, such as those with hydrogen as C-H (aliphatic compounds, fat, oils, proteins), N-H (proteins) or O-H (water, alcohols, acids), is especially useful to determine the composition of organic matter. This makes this technique suitable for the quantification of the chemical composition of samples with a high percentage of fat, protein or water content and allows the determination of other parameters that depend on many chemical and physical properties such as the species [26], freshness, storage conditions (unfrozen or frozen-thawed fish) [27] or even non-polar analytes in low concentrations through indirect ways.

For all the previously mentioned reasons, NIRS, coupled with chemometrics, is starting to be considered as essential as other analytical methods for fish analysis. Indeed, the number of Web of Science indexed articles in which NIRS techniques have been applied to determine fish quality, safety or authenticity have increased rapidly during the last 20 years. While this technique is mostly used for fish quality control and authenticity, applications such as fish safety have begun to be explored in the last 10 years (**Figure 3**).

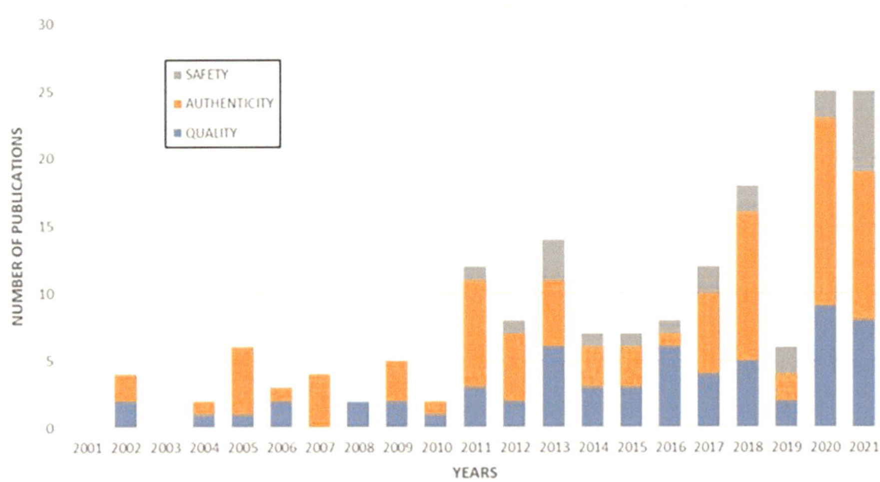

Figure 3.
Number of publications since 2001 regarding NIR and fish. Data extracted from Web of Science in April 2022. The keywords used in topic, common for all the sections, were 'fish' AND 'NIR' OR 'near infrared spectroscopy'. For quality control, two independent searches were made: i) 'composition' OR 'fat protein' AND 'quality' ii) 'freshness' AND 'storage' NOT 'safety'. For authenticity, four independent searches were made: i) 'specie' AND 'identification' OR 'substitution' OR 'authentication' ii) 'origin' iii) 'farm* wild' OR 'production method*' iv) 'fresh thawed' OR 'unfrozen thawed' OR 'fresh frozen' OR 'unfrozen frozen'. For safety, two independent searches were made: i) 'spoilage' OR 'toxic' OR '*amine*' AND 'safety' ii) 'micro*' OR 'bacteria' AND 'safety'.*

2. Applications of NIRS in fish chain

While writing this chapter, the authors reviewed the use of NIRS in the fish chain regarding three different application fields: fish quality control, fish authenticity and fish safety (**Figure 4**). The supplementary material, which shows a summary of the works that will be reviewed in this chapter, is available in the **Appendix**.

2.1 Fish quality control

Assuring the quality of products is one of the objectives that any industry must consider, and it is of major importance in the fish chain, due to the problems that are usually associated with its evaluation at an industrial level [28]. Unlike other foodstuff, quality of fish and fish products is more difficult to control, due to several factors, such as the variations in species, sex, age or habitats [29]. In this sense, fish quality control refers specially to the determination of the proximate composition of the fish samples and the evaluation of the freshness of the products. From one side, the measurement of the proximate composition of fishery species is very important, since it varies from one species and one individual to another. Moreover, it also gives an idea of the sexual stage of the fish [30], and it changes depending on many factors (diet, origin, rearing systems). The proximate composition determines the organoleptic quality of a specific fish product [31]. At the same time, the determination of the freshness of samples is another indicator of the quality of the product. Freshness is an ambiguous term, which can be interpreted in different ways, and it should not be used unless it is properly defined. Through this chapter, the concept of freshness is going to be a quality descriptor of fish samples, and it will refer to the time past after the catch of the fish. If that time is short, the sample will be considered fresh, as it will retain its original characteristics. As time passes, samples will lose quality, due to the action of biochemical, chemical, physical and microbial processes, and therefore losing freshness [32]. Nowadays, NIRS is becoming an alternative as a quality control method, due to its advantages over traditional analysis [33].

2.1.1 Proximate composition

Fish is mainly composed of water, protein, fat and other minoritarian compounds [28]. As fish proximate composition is a quantitative analysis, regression techniques (usually partial least square regression or PLSR) are used to determine the composition of fish by NIRS, as an alternative to laboratory determinations.

Many authors have determined the proximate composition of different fish species using NIRS. In one of the first studies, NIRS was used to determine moisture, fat and

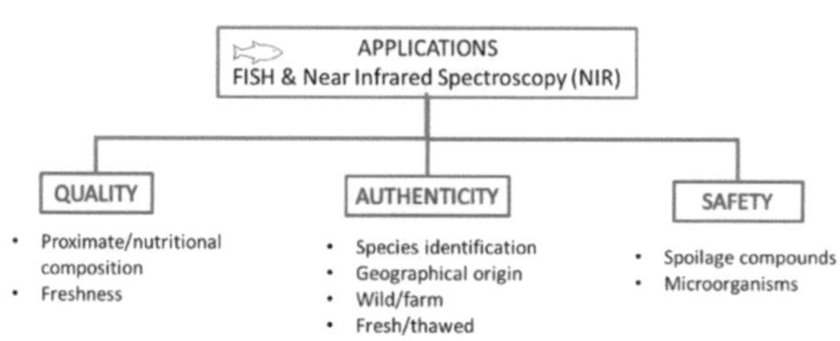

Figure 4.
Classification of the problematics considered and reviewed under this chapter.

protein content in Atlantic salmon (*Salmo salar*) fillets and in minced samples [34]. Principal component regressions (PCRs) were used to perform the calibrations, which were evaluated using a full cross-validation. This study illustrated the potentiality of NIRS for predicting the proximate composition of minced fish, obtaining with the best models a root mean square error of cross-validation (RMSECV) of 6.6, 3.8 and 3.0 g/Kg in fat, moisture and protein analysis, respectively. However, they failed in the analysis of the whole fish, with RMSECV higher than 13.2 g/Kg in all the models. Years later, another study performed PLSR directly in fish fillets of Chinese export tilapia [31]. The regressions showed good results for the determination of moisture and lipid content, with r^2 = 0.96 and 0.97 and standard error of cross-validation or SECV = 0.87 and 0.71, respectively. Even though the regression made on the protein content had a lower coefficient of determination in the cross-validation (r^2 = 0.31), it showed good results regarding the error (SECV = 0.55). This difference was due to the small protein differences that tilapia fillets have between them, and it could be solved adding samples with more variability to the calibration set. It is important to highlight that they performed an external validation and not only a cross-validation, for moisture and protein determination, using new model-outside samples, which presented variations that authors considered acceptable between predicted and measured values (a maximum relative error of 1.67% for moisture and 6.48% for lipid). Similar studies regarding the analysis of the proximate composition in fish samples are shown in **Table 1**.

2.1.2 Freshness

In this chapter, freshness has been considered as a quality characteristic, and it will be defined as the time passed after the catch. It is also influenced by the temperature during the storage [32]. Freshness is usually evaluated using sensory analyses, which consider several parameters such as flesh elasticity, pupil colour and skin odour, among others [43].

Although principal component analysis (PCA) is not a classificatory nor a discriminant method (since it is a non-supervised technique, which only takes into account the spectral information), and it should not be used with that purpose, some authors have applied it in order to see the differences between fresh fish and fish samples that have been stored in the fridge for days. In this sense, some authors used short-wavelength near infrared spectroscopy (SW-NIRS) to differentiate fresh from non-fresh rainbow trout (*Oncorhynchus mykiss*) samples, in fillets and minced [44]. Fish samples were stored at 4°C for 8 days, or at room temperature (21°C) for 24 hours. PCA results showed a clear separation between control samples (1 day stored) and fish kept 4 days or longer in refrigeration. Regarding fish at room temperature, a segregation was found for samples stored 10 hours or longer in comparison to control samples (stored 0 hours). Other studies performed PCA to discriminate between fresh Atlantic salmon (*Salmo salar*) and salmon stored at 4°C for 9 days, finding a clear separation between fresh and non-fresh fillets [45]. A similar experiment used PCA to discriminate golden pompano (*Trachinotus ovatus*) samples regarding the storage time using visible/near-infrared spectroscopy (VIS/NIRS) [46]. However, they concluded that samples stored for 0, 2 and 3 days could not be well discriminated using this analysis. They also created models to predict the storage time, but the model which used VIS/NIRS spectroscopy was not validated (only a model combining VIS/NIRS data and electronic nose information was validated, obtaining an accuracy of 93.3%). Another study analysed the spectra of rohu fish (*Labeo rohita*) sliced into pieces, using SW-NIRS and PCA to distinguish between

Fish species	Sample presentation	Measured parameters	Algorithm	Results	Reference
Atlantic salmon (*Salmo salar*)	Whole	Fat and moisture	PLSR	r^2 = 0.87, 0.86; RMSECV = 1.12, 0.98%, respectively	[35]
Sea bass (*Dicentrarchus labrax*)	Fillet portions, whole fresh minced fillet, freeze-dried minced fillet	Water, ether extract and crude protein	PLSR	SECV <2.81% in all the cases	[36]
Sea bass (*Dicentrarchus labrax*)	Fresh and freeze-dried minced fillets	Water, ether extract and crude protein	PLSR	SECV ≤0.39% in all the cases	[37]
Skipjack tuna (*Katsuwonus pelamis*) and yellow fin (*Thunnus albacares*)	Fish flesh	Moisture, protein, free and total fat	PLS-2	r^2 > 0.90 in all, small deviations between reference and predicted values in the cross-validation	[38]
Nile tilapia's (*Oreochromis niloticus*)	Fillets	Fat content and FAs	PLSR	correlation coefficient = 0.70, 0.71; (RMSEP) = 2.39, 4.76 mg FA/g of total lipids, respectively	[39]
Saithe (*Pollachius virens*) and hoki (*Macruronus novaezelandiae*)	Thawed fish pieces	Total lipid content, phospholipids, (PUFAs)* and (MUFAs)	PLSR	r^2 ≥ 0.73, 0.96. The highest RMSEP = 5.31, 4.10%, respectively.	[33]
Cured salmon roe (*Oncorhynchus keta*), cured Atlantic salmon (*Salmo salar*) and cold smoked Atlantic salmon (*S. salar*).	Roe and salmon pieces	Moisture and sodium chloride	PCR, MLS, PLSR and different neural network algorithms	r^2 ≥ 0.71 and standard error ≤ 1.87%, r^2 ≥ 0.72 and standard error ≤ 2.04%, r^2 ≥ 0.64 and standard error ≤ 4.10% in the cross-validation for both components, respectively	[40–42]

**PUFAs: polyunsaturated fatty acids; MUFAs: monounsaturated fatty acids, FAs: fatty acids; MLR multiple linear regression; RMSEP: root mean squared error of prediction.*

Table 1.
Use of NIRS to analyse proximate composition of fish.

fresh fish (after being collected) and fish stored in the fridge at 4°C during 4, 7 and 12 days [47]. In the PCA analysis, samples were well grouped by their freshness, so that four clusters were created (one per each measurement day).

Different studies aim not only to separate fresh samples from non-fresh ones, but also to make models to correlate spectral data with the storage time. One of them explored the suitability of SW-NIRS to estimate the freshness of the fillets of two fish species: cod (*Gadus morhua*) and salmon (*Salmo salar*) [32]. PLSR models were created to predict the number of days that the fish has been stored and validated using a full cross-validation. Cod gave the best correlation with the visible wavelength range (r^2 = 0.97, RMSE = 1.04 days) and salmon with the NIR range (correlation of 0.98 and RMSE = 1.20 days). A similar study predicted the freshness in cod (*Gadus morhua*) fillets with a handheld interactance probe, using VIS/NIRS. Fillets were stored in a cold room (between 2 and 4°C) for 13 days, making measurements at days 0, 2, 4, 7, 10 and 13 [48]. PLSR models were validated using a leave-one-out cross-validation. The best results were obtained selecting some wavelengths of the visible range (448, 487, 606 and 646 nm) and using SNV, giving a correlation coefficient of 0.93 and an RMSECV = 1.66 days. Other researchers used VIS/NIRS to perform a similar study on Atlantic salmon (*Salmo salar* L.) fillets [49]. In those, PLSR models using the wavelength region between 605 and 965 nm and the pre-treatment with SNV gave the lowest error in the leave-one-out cross-validation (RMSECV = 1.91 days). In another work the freshness of cod (*Gadus morhua*) was predicted using VIS/NIRS [50]. Authors built PLSR models to predict the quality index method (QIM), a sensory analysis made to determine the freshness of cod fillets. A full cross-validation was used to test the performance of the models, giving the analysis made with the visible part of the spectra the best results (correlation coefficients = 0.96 and RMSE = 2.6 points).

2.2 Fish authenticity

The globalization and expansion of the fish and aquaculture sector, in addition to the increasing public concern about food quality, have caused a growing interest in several issues related to fish authenticity. According to the European Regulation (EU) n. 1379/2013 [51], fishery and aquaculture products must be labelled with the commercial designation, proper scientific name of the species, production method (e.g. caught, farmed), fishing gear (e.g. hook, trap, trawl), catch or production area and storage method (unfrozen or frozen-thawed). Other claims, such as environmental and production techniques, can also be reported on the label, and all this information must be verified by effective methods. Special mention is the differentiation of farmed from wild-produced seafood, which is becoming a concern among consumers, since the quality and the price may be affected by the production method. The presence of errors in the label information about fish origin and production process is increasing, so that nowadays fish is the second most vulnerable category of food to fraud [52]. In order to protect consumers and with the aim to avoid it, the assessment of seafood origin is increasing as a security measure. Regulatory interventions are trying to stop the mislabelling or the substitution of wild with farmed fish, which is mitigating risks for the consumer's confidence and health [53].

Both the geographical origin and the production method, among others, can strongly affect the characteristics of the two types of products, whose discriminating properties are usually difficult to determine. Several analytical techniques have been traditionally used to assess fish authenticity. However, even though they are well established, there is still a necessity for faster, easier and more affordable methods [52].

2.2.1 Species identification

The substitution of valuable species with cheaper ones at any point of the supply chain is one of the most common frauds, sometimes in an unintentional way due to the similarities between species or the use of different names for the same species. However, the differences in the economic value, the exploitation of endangered species, the replacement with poisonous fish and their difficult identification, make the substitution of fish species an extended problem, especially severe after processing, at the retailers and supermarkets [54]. The mentioned frauds particularly affect fish fillets and ready-to-eat products, such as fish patties, which cannot be recognized through the traditional morphological analysis.

The use of NIRS with the objective of discriminating between species has been explored during the last years [52]. For instance, in a preliminary study under industrial conditions, some authors used NIRS coupled with partial least square discriminant analysis (PLS-DA) and linear discriminant analysis (LDA) on PCA scores to discriminate between fishmeal made with different fish species: salmon (*Salmo salar*), blue whiting (*Micromesistius poutassau*), mackerel (*Scomber scombrus*) or herring (*Clupea harengus*) [55]. They obtained a correct prediction percentage > 80%, tested using cross-validation.

In other studies, authors measured the skin and meat of two different species of cod (winter cod and cod), mullet (red mullet and Atlantic mullet) and trout (samlet and salmon trout) with a handheld NIR spectrometer [56]. The data were analysed using PCA and soft independent model of class analogies (SIMCA), obtaining promising results with the last method. Other authors used Fourier transform NIR (FT-NIR) to distinguish fillets of red mullet (*Mullus surmuletus)* and plaice (*Pleuronectes platessa*), fish with higher economic value, from Atlantic mullet (*Pseudupeneus prayensis*) and flounder (*Platichthys flesus flesus*), species which are cheaper [57]. LDA and SIMCA were used to classify them, obtaining a sensitivity higher than 70% and a specificity of 100% for FT-NIR and SIMCA, and a 100% correct classification with LDA for both techniques.

The differentiation between different carp species: silver (*Hypophthalmichthys molitrix)* black (*Mylopharyngodon piceus*), bighead (*Aristichthys nobilis*), grass (*Ctenopharyngodon idellus*) and common carp (*Cyprinus carpio*) and also crucian (*Carassius auratus*) and bream (*Parabramis pekinensis*) has also been studied [58]. Only nine selected wavelengths in the 1000–1799 nm region after the pre-treatment were used. In this case, two models using PCA-LDA and the fast Fourier transformation pre-processing (FFT) before LDA analysis were built. They showed a 100% accuracy, specificity, sensitivity and precision. Fish surimi made from white croaker (*Argyrosomus argentatus*), hairtail (*Trichiurus haumela*) and redcoat (*Nemipterus virgatus*) was successfully classified by using LDA and multiplicative scatter correction (MSC) as pre-treatment. The results of the prediction of the calibration and the validation data sets were 98.5% and 100%, respectively [59].

More recently, an attempt to distinguish Atlantic cod (*Gadus morhua*) from haddock (*Melanogrammus aeglefinus*) was made. Raw fillets and patties were measured with a handheld and a benchtop instrument, with the spectra pre-processed using SNV, MSC or SG, and analysed with LDA and SIMCA. LDA with MSC pre-treatment obtained a 100% of correct classifications for both fillets and patties for the external validation set, regardless of the instrument. The SIMCA method also obtained a 100% of successful classification of fillets using SNV and SG (first and second derivative) for the benchtop instrument and using MSC for the portable NIR device. With SIMCA, 100% correct predictions of patties were obtained for all instruments and pre-processing techniques [60].

2.2.2 Farmed vs. wild fish and types of farming

In the last years, there has been an increasing interest from farming industries, retailers and even from consumers in investigating the differences between cultured and wild fish [61]. Most of the studies which differentiate farm from wild fish are centred in fish species such as European sea bass (*Dicentrarchus labrax* L.), which is one of the most economically important fish species in the whole Mediterranean area [62]. The intense competition among producing countries in the Mediterranean Sea (Turkey and Greece are the largest producing countries and Italy, Spain and France are the world's leading importers) and the lowering of the market prices are requiring ways to differentiate sea bass quality. The differences in the production methods (farm and wild specimens) and, among farmed individuals, in the rearing systems and the feeding regimens used, may affect the flesh quality. Usually, farmed fish presents lighter fillet colour, softer texture and milder flavour than wild fish [36, 61]. Previous studies aimed to differentiate between farmed and wild sea bass were based on discriminating the two classes depending on their protein and amino acids composition, metallic ions and fatty acids groups, being usually the lipid profile the differential factor. However, the extraction of lipids and the transformation of the triglycerides into their corresponding fatty acid methyl esters (FAMEs) plus the analysis by chromatography with a flame ionization detector (GC-FID) are a tedious and time-consuming methodology, which not only generates toxic waste but can also provoke the oxidation of the polyunsaturated groups [63].

NIRS has shown its potentiality in some studies as a rapid method to classify sea bass according to rearing systems [53]. Some authors used an NIR spectrometer to discriminate between four types of sea bass rearing systems: extensive, semi-intensive, intensive and sea-cage in fresh minced fish and freeze-dried minced samples [36]. They performed a four-cluster SIMCA analysis, which was evaluated using a full cross-validation. The best results were obtained with the freeze-dried minced fillets (83%, 80%, 74% and 83% of samples well classified for extensive, semi-intensive, intensive and sea-cage methods, respectively). Similar results were obtained by [37] in the discrimination of organic or semi-intensive sea bass. Some years later, NIRS was also used to discriminate between farm and wild seabass fillets, which were minced before the analysis [53]. They developed three different approaches: an authentication using measured chemical variables (PLS-DA_mc), an authentication using estimated chemical variables (PLS-DA_ec), an authentication with direct use of spectral data (PLS-DA_NIR) and the WPTER algorithm. The evaluation of the validation data set gave very similar results for all the techniques in the classification of declared wild samples. Regarding the declared farming samples, PLS-DA_mc classified correctly all the samples, while PLS-DA_ec misclassified 2, and PLS-DA_NIR and WPTER misclassified 3 (in all the cases out of 34). All the methods indicated that the most relevant spectral regions in the models were the ones related to the absorbance of the CH, CH_2, CH_3 and H_2O groups, which corresponded to fat, fatty acids, and water content, highlighting that the difference between farm and wild sea bass is very related with the fat/water percentage and the lipid profile. Other authors also studied the differences between sea bass, according to the production method (wild or farmed) and the rearing system used (extensive, semi-intensive or intensive) [62]. In this study, an NIR spectrometer was used to analyse minced and homogenized fillets. In this case, cross-validation and an external validation were made. PCA was used as exploratory analysis and two different orthogonal partial least square-discriminant analysis (OPLS-DA) models were performed. The evaluation of the external validation showed an overall classification rate of a 100% for the production method and 94.44% for the farming system. Looking at the categories, all the farm and wild samples were correctly classified, while regarding the farming system, wild samples presented a classification rate of 100%, extensive samples 66.67%, semi-intensive samples 80.00% and intensive samples

100%. Although most of the studies regarding discrimination between farm and wild fish are performed in sea bass, other species have also been analysed. Segregation between wild and farm sole (*Solea solea*) and turbot (*Psetta maxima*) has also been performed [64]. Here, several chemometric methods were tested, obtaining the best results with logistic regression in sole (90% of precision, recall and F1-score) and with support vector machine in turbot (99% of precision, recall and F1-score) in the evaluation of the test set.

2.2.3 Geographical origin

NIRS is one of the most used techniques to determine the origin of a lot of types of foods (oil, coffee, wine, etc). However, especially in fish and fish products, its use for such purpose is a challenge due to the difficulty in proving the geographical origin of a particular fish. A very large number of samples must be measured for each location due to a high heterogeneity among them, depending on the season or different years, the available food, if the fish has been captured before or after the spawning season, etc. With the aim of determining the viability of NIRS regarding the geographical origin, it often requires multi-disciplinary methods that could consider all the environmental and genetic factors that determine fish characteristics. Even so, there are some studies that successfully applied NIRS to determine the geographical origin of fish products.

Some authors used NIRS to classify tilapia fillets in four different Chinese regions (Guangdong, Hainan, Guanxi and Fujian provinces) [31]. The classification models were carried out with SIMCA, achieving more than 80% correct classification for the Guangdong, Hainan and Fujian and 75% of the Guangxi samples by the cross-validation method. None of them were assigned to more than one category, less than 20% to none of the classes and only less than 2% were incorrectly classified. The differentiation between minced fish fillets made using European seabass (*Dicentrarchus labrax* L.) caught in three areas of the Mediterranean Sea has been also explored [62]. In this case, NIRS was used in combination with PLS-DA method. The external validation with a test set resulted in 100% of the eastern, 88% central and 85% western Mediterranean samples were correctly classified. In another study, NIRS was used successfully to assess the geographical origin (Morocco, Spain, Tunisia or Croatia) of minced semi-finished and finished salted anchovies (*Engraulis encrasicholus*) [65]. The combination of models based on orthogonal PLS-DA gave the correct identification of the origin of both types of anchovies with >98% sensitivity, > 99% specificity and > 99% accuracy in an external validation using a test set with the 20% of the samples. Moroccan samples were distinguished for the proteins and degradation compounds absorption bands, Tunisian anchovies for the unsaturated lipids, while Croatian and Spanish samples were classified for their differences on both types of absorption bands.

2.2.4 Unfrozen vs. frozen-thawed fish

Freezing, due to its advantages, is intensively used for preserving fish. It maintains the product safety, the sensory qualities and also the nutritional characteristics. However, freezing the products, as well as the storage and the thawing process, changes the physical and biochemical properties of the fish muscle [66]. Recent technological advances in freezing and the posterior storage have allowed to minimize the damage caused by the thawing process, while maintaining the product quality. The impact the freezing process does on the product depends on several factors inherent to fish characteristics (genetics, muscle type, husbandry) and the post-mortem treatments (freezing rate, storage time, storage pressure and temperature, thawing method) [67]. Slow freezing leads to the formation of large extracellular ice crystals which damage the cell membranes and muscle proteins [68–70]. Also, in general, freezing-thawing cycles are related to the degradation of textural properties due to physical damages and protein denaturation and oxidation [66].

As it has been reported before, most of the analytical techniques used for determining whether the fish has been frozen or not, despite their accuracy, are destructive, time-consuming and expensive, making them not suitable for commercial applications [71]. As an alternative, NIRS, in combination with qualitative chemometric methods, has been adopted as an efficient solution for this determination.

Several authors have used NIRS and chemometrics to discriminate between unfrozen and frozen-thawed fish. FT-NIRS has been used, coupled with two different classification algorithms (LDA and SIMCA), to discriminate between unfrozen and thawed Atlantic mullet (*Pseudupeneus prayensis*) fillets [57]. LDA provided the most accurate results in the evaluation of the external validation set, with correct classification rates of 97.2% for unfrozen and 100% for thawed samples. The same classification algorithms were used in [72] to distinguish whether red sea bream (*Pagrus major*) samples were unfrozen or frozen-thawed using an NIR spectrophotometer in interactance mode. They observed that the change in total reflectance, the major effect of the freeze-thawing process, arises from changes in the scattered light at the surface of the fish due to alterations in its structure. The classification accuracy when using absorbance spectra without any pre-processing (100%) was much higher than for spectra pre-treated with the multiplicative scatter correction (MSC) (81%), confirming that light scattering contains most of the information. VIS/NIRS techniques were as well used to distinguish unfrozen from thawed swordfish (*Xiphias gladius L*) cutlets stored at different temperatures (−10 and − 18°C) for several days [73]. VIS/NIR spectra were collected using a portable spectrophotometer and a NIR monochromator. The correct classification percentage for VIS/NIR spectra was ≥96.7%, while for the NIR spectra was ≥90.0%, with no effect on the classification due to the temperature or days that the frozen samples were stored. In another reference, the authors distinguished unfrozen from frozen tuna fillets (*Thunnus thynnus*) which were frozen and, after 5, 21 and 35 days, thawed and then re-scanned [74]. In this study, the PLS-DA algorithm with double cross-validation was used to classify unfrozen from frozen-thawed fillets with a 92% and 82% of correctly classified samples, respectively. Some researchers have tried to discriminate between tilapia fillets (*Oreochromis*) that have been subjected to different freezing-thawing cycles, from one to seven [75]. In this project, NIR spectra were collected at each cycle and the test data analysed with PCA and discrimination analysis, obtaining a correct classification of 86% for once frozen-thawed fillets and 93% for samples subjected to several freezing cycles. Finally, in other recent study, unfrozen and frozen-thawed bigeye tuna (*Thunnus obesus*) fillets were also measured with NIRS for their discrimination [76]. To include variability, samples were injected with different commercially used concentrated solutions (NaCl, polyphosphates and protein hydrolysate). Three different models, based on PLS-DA, were developed giving NIRS the best results with an accuracy of 0.91 and an error rate of 0.10 for the prediction of the validation set.

2.3 Fish safety

Fish and fish products are very perishable, being subject to a fast deterioration process and becoming easily unsuitable for consumption and, therefore, a possible threat to public health. The biological composition and characteristics of fish, i.e. low percentage of connective tissue, autolytic enzymes, neutral pH, high water activity [27] and the growth of spoilage bacteria, contribute to the short shelf-life of these products [45]. It is estimated that 25% of the fish losses are initiated by microbial and chemical deterioration [77].

Even though most studies regarding NIRS in the fish industry have been conducted to determine their quality and authenticity, the use of this technique in food safety evaluation is still relatively new but constantly increasing [25]. During the last years, many studies have been conducted in the field of fish safety, in order to determine the presence of spoilage compounds and microorganism in fish products.

2.3.1 Detection of spoilage compounds

Some toxic compounds produced during fish spoilage can be measured by NIRS to evaluate fish safety. An important group is formed by biogenic amines, with histamine being one of the most critical, as it is related with the most common foodborne illness associated with fish consumption [78]. They are a group of nitrogenous compounds, usually measured to evaluate the safety of aquatic products. Among them, the most common are histamine, putrescine, cadaverine and tyramine. Biogenic amines are formed by the decarboxylation of amino acids or by the amination and transamination of aldehydes and ketones. The consumption of fish products with excessive levels of biogenic amines may lead to food intoxication, so they have become an important index to evaluate fish safety [79].

NIRS has been used as a tool to determine the cadaverine content in homogenized and ultrafiltered solutions of Chub Mackerel (*Scomber japonicus*) stored at different temperatures (5 and 25°C) [80]. The r^2 obtained in the regression analysis was 0.98 and 0.99 for the different dilutions made. However, authors mentioned that NIRS measurements of cadaverine content made directly in the fish were not successful. Histamine was also determined by this spectroscopic technique. It is the only biogenic amine for which international regulatory levels have been established [78]. The presence of this amine is due to the bacterial decarboxylation of histidine, an amino acid present in high levels in fish muscle, specially from Scomberiscida and Scombridae families (which include tuna). Studies have tried to predict this compound in order to assure fish safety. Histamine content has been measured in different dry extracts of skipjack tuna (*Katsuwonus pelmis*) using NIRS [81]. The standard error of prediction or SEP in all the cases (using different solvents and different pre-processing techniques) was between 2.94 and 3.47 in the validation set, with r^2 between 0.63 and 0.79. Later on, in a similar experiment, histamine in raw (*Thunnus albacares*) and in processed tuna (*Katsuwonus pelamis* and *Thunnus albacares*) was detected, which were previously minced and homogenized [78]. Both models showed r^2 values higher than 0.97 and RMSEP lower than 10 mg kg^{-1}, with the obtained results being better for processed tuna. Considering the mentioned studies, promising results have been seen in all the works regarding the detection of biogenic amines in fish by NIRS. However, more efforts are still needed to determine these compounds directly in the fish fillet, without the necessity of destroying the sample.

Biogenic amines are not the only compounds associated with the spoilage of fish. Trimethylamine concentration can also be used as an indicator of fish spoilage, since it is related with the putrefaction process and the fishy odour. This compound has been estimated in fresh fish (*Hypophthalmichthys molitrix*) using FT-NIRS and different chemometric algorithms [77]. In this case, measurements were made directly in the fish, which was sliced in small cubes. Regression analysis showed RMSEP between 5.10 and 5.75 mg N/100 g and correlation coefficients of prediction between 0.94 and 0.98. K-value is another indicator of fish deterioration that is related to the degradation of ATP in the fish flesh. Likewise, this index was also estimated in fresh silver carps (*Hypophthalmichthys molitrix*) using FT-NIRS to make the measurements directly in the fish. Several chemometric techniques were compared, obtaining as best result an RMSEP of 0.036 and a correlation coefficient of prediction of 93.74% [82]. Another way to measure the spoilage of saithe (*Pollachius virens*) and hoki (*Macruronus novaezelandiae*) fillets has been also explored [33]. Here the authors developed PLSR models to predict free fatty acids (FFA), fluorescent interaction compounds (OFR) and thiobarbituric acid reactive substances (TBARS), in order to measure the lipid oxidation, one of the major deteriorative reactions in fish muscle. Results showed acceptable results for the test set in both species (r^2 between 0.63 and 0.98), although some high errors were found (RMSEP of 2.90 and 3.43 g FFA/100 g lipids, 0.97 and 0.55 μmol malonaldehyde diethylacetal/kg of samples for TBARS and 5.23 and 1.20 for OFR in hoki and saithe, respectively). Similar results were found regarding FFA [83] and on finding lipid oxidation products in vegetable oils [84]. Other similar analysis were performed using NIRS to predict the pH, TBARS, total volatile basic nitrogen (TVB-N) and K-value of minced bighead carp (*Aristichthys nobilis*) [85]. PLSR together with competitive adaptive reweighted sampling (CARS) algorithm and pre-processing methods were used to make regressions, obtaining satisfactory results for the prediction of the test set (coefficients of prediction of 0.945, 0.932, 0.954 and 0.807 and RMSEP of 0.081, 2.099, 0.107 and 6.509 for pH, TVB-N, TBA and K value, respectively).

2.3.2 Microorganisms

Fish is one of the most vulnerable aquatic products, as it serves as a growth medium for microorganisms, which can be either pathogenic or cause fish spoilage. In this sense, spectroscopic techniques have shown potential for spoilage monitoring to confirm if fish is safe for consumption or not, since conventional methods to determine microorganisms are really time-consuming and tedious [86]. However, although some works have been done regarding the detection of microorganisms by NIRS and the potentiality of the technique is big, this is an application that must be studied carefully, and it must be validated using techniques with limits of detection more sensitive than NIR.

Some studies tried to use SW-NIRS to quantify microbial loads (expressed as total viable count) in rainbow trout (*Oncorhynchus mykiss*) [44]. Regression models were built based on spectra taken in the flesh and skin of fish fillets, stored at 4°C for 8 days, and also in minced samples stored at room temperature 24 hours. All the models gave goods results, the model made in the flesh being slightly better (r^2 = 0.97 and SEP = 0.38 log CFU/g) compared with the models of the skin and the minced samples (r^2 = 0.94 and 0.82 and SEP = 0.53 and 0.82 log CFU/g, respectively). Similar experiments have used FT-NIRS to detect and predict microbial spoilage on Atlantic salmon fillets (*Salmo salar*) [45]. The results showed that the regression analysis was able to predict the numbers of bacteria (expressed as total aerobic plate counts) after 9 days of storage at 4°C, with an r^2 of 0.64 and a RMSEP of 0.32 log cfu/g in the validation set. A portable spectrometer was also used to determine the total bacteria content in flounder fillets (*Paralicthys olivaceus*), expressed as the aerobic plate count of total bacteria, after 8 days of storage at 4°C [87]. Different models based on different chemometrics techniques were developed, obtaining the best results using a combination of genetic algorithm and artificial neural networks, with an r^2 of 0.966 and an RMSEP of 0.083 in the test set.

3. Conclusions

NIRS has been proved as an effective analysis tool for the fish value chain. Although the use of this technology in fish and fish products routine inspection has recently increased, more and more studies are available each year, some applications are more developed than others. The monitoring of the quality and authenticity of the fish chain by NIRS is well known, with applications that are nowadays implemented in the food industry. However, more efforts are still needed regarding the use of this technology for safety issues, where the limit of detection still represents a big challenge. Even if some studies have been carried out successfully during the last decade, these applications should be studied carefully, and more investigations must be done before trying to implement NIRS as an alternative to the already established destructive methods.

Acknowledgements

The authors greatly acknowledge the Basque Government—Department of Economic Development, Sustainability and Environment—Vice. Dept. of Agriculture, Fishing and Food Policy, Directorate of Quality and Food Industries for the funding of the project ELIKatea 4.0 and for the scholarship of Sonia Nieto-Ortega.

Appendix

Application	Species (sample preparation)	Measured property	Instrument	Chemometric method	Statistical results	Reference
Proximate composition (Quality)	Tilapia (fillets)	Moisture, protein and lipid	Reflectance NIR	PLSR	r^2cv = 0.959, SECV = 0.872 g/100 g for moisture r^2cv = 0.306, SECV = 0.550 g/100 g for protein r^2cv = 0.971, SECV = 0.706 g/100 g for lipid	[31]
Geographical origin (Authenticity)		Chinese region origin		SIMCA	> 75% Correct classification	
Freshness (Quality)	Cod *Gadus morhua* Salmon *Salmo salar (fillets)*	Storage time	Transflection Vis–NIR	PLSR	r^2p = 0.97, RMSEP = 1.04 days for cod in visible region r^2p = 0.98, RMSEP = 1.20 days for salmon in NIR region	[32]
Proximate composition (Quality)	Saithe *Pollachius virens* Hoki *Macruronus novaezelandiae* (thawed portions)	Total lipid content, phospholipids, PUFAs and monounsaturated fatty acids (MUFAs)	Reflectance FT-NIR	PLSR	$r^2v \geq 0.96$ hoki and $r^2v \geq 0.73$ saithe for lipid composition RMSEP <4.10 hoki and RMSEP <5.31% saithe for lipid composition	[33]
Spoilage compounds (Safety)		Lipid oxidation components: free fatty acids (FFA), fluorescent interaction compounds (OFR) and thiobarbituric acid reactive substances (TBARS)		PLSR	$r^2v \geq 0.63$ hoki and $r^2v \geq 0.67$ saithe for lipid oxidation RMSEP <5.23 hoki and RMSEP <3.43% saithe for lipid oxidation	
Proximate composition (Quality)	Atlantic salmon *S. salar* (fillets and minced)	Moisture, fat and protein	On-line NIR fibre-optic probe and off-line reflectance NIR	PCR	RMSECV = 6.6 g/Kg for fat in minced sample RMSECV = 3.8 g/Kg for moisture in minced sample RMSECV = 2.0 g/Kg for protein in minced sample RMSECV = 12.4 g/Kg for fat in whole fillet RMSECV = 11.3 g/Kg for moisture in whole fillet	[34]

Application	Species (sample preparation)	Measured property	Instrument	Chemometric method	Statistical results	Reference
Proximate composition (Quality)	Atlantic salmon salmo salar (whole fish)	Fat and moisture content	Interactance NIR	PLSR	$r^2cv = 0.87$, RMSECV = 1.12% for fat $r^2cv = 0.86$, RMSECV = 0.98% for moisture	[35]
Proximate composition (Quality)	European sea bass *Dicentrarchus labrax* (fillet, fresh minced and freeze-dried minced)	Water, ether extract and crude protein	Absorbance NIR + monocromator	PLSR	$r^2cv > 0.47$, SECV <2.76% for water in fillets $r^2cv = 0.95$, SECV = 0.87% for water in minced $r^2cv = 0.97$, SECV = 0.70% for water in freeze-dried $r^2cv > 0.48$, SECV <2.81% for ether in fillets $r^2cv = 0.97$, SECV = 0.70% for ether in minced $r^2cv = 0.97$, SECV = 0.62% for ether in freeze-dried $r^2cv = 0.30$, SECV = 0.55% for protein in minced $r^2cv = 0.68$, SECV = 0.35% for protein in freeze-dried	[36]
Rearing system (Authenticity)		Extensive, semi-extensive, intensive or sea-cages		SIMCA	> 37% correct classification for minced fillets > 74% correct classification for freeze-dried minced fillets	
Proximate composition (Quality)	European sea bass *Dicentrarchus labrax* (fresh and freeze-dried minced)	Water, ether extract and crude protein	Absorbance NIR + monocromator	PLSR	$r^2cv = 0.935$, SECV = 0.387% for water in minced $r^2cv = 0.953$, SECV = 0.328% for water in freeze-dried $r^2cv = 0.984$, SECV = 0.154% for ether in minced $r^2cv = 0.982$, SECV = 0.175% for ether in freeze-dried $r^2cv = 0.587$, SECV = 0.290% for protein in minced $r^2cv = 0.640$, SECV = 0.274% for protein in freeze-dried	[37]

Application	Species (sample preparation)	Measured property	Instrument	Chemometric method	Statistical results	Reference
Rearing system (Authenticity)		Organic or semi-intensive		SIMCA	> 20% correct classification for minced samples >65% correct classification for freeze-dried samples	
Proximate composition (Quality)	Skipjack tuna *Katsuwonus pelamis* and yellow fin tuna *Thunnus albacares* (cutlets)	Moisture, protein, free and total fat	NIR + fibre optic probe	PLS-2	r^2p = 0.98 for moisture r^2p = 0.99 for protein r^2p = 0.95 for total fat r^2p = 0.96 for free fat	[38]
Proximate composition (Quality)	Tilapia *Oreochromis niloticus (fillets)*	Omega-3 and omega-6	Reflectance NIR	PLSR	r^2p = 0.70, RMSEP = 2.39 mg FA/g for omega-3 r^2p = 0.71, RMSEP = 4.76 mg FA/g for omega-6	[39]
Proximate composition (Quality)	Salmon roe *Oncorhynchus keta* (cured)	Salt and moisture	SW-NIR + fibre optic probe	PLSR	r^2p > 0.711, SEP < 1.13% for salt r^2p > 0.609, SEP < 1.87% for moisture	[40]
Proximate composition (Quality)	Atlantic salmon *Salmo salar* (cold smoked)	Salt, water and water phase salt (WPS)	SW-NIR + fibre optic probe	PCR, MLR, PLSR, BPNN	r^2cv > 0.639, RMSCV <0.77 for salt r^2cv > 0.854, RMSCV <4.10 for water r^2cv > 0.717, RMSCV <1.55 for WPS	[41]
Proximate composition (Quality)	Atlantic salmon *Salmo salar* (cured portions)	Salt and moisture	SW-NIR + fibre optic probe	PLSR, BPNN	r^2cv > 0.701, RMSCV <1.43 for salt r^2cv > 0.784, RMSCV <2.08 for moisture	[42]
Freshness (Quality)	Rainbow trout *Oncorhynchus mykiss* (flesh and skin of fillets and minced)	Storage time (2 temperatures)	SW-NIRS	PCA	Segregation between fresh and non-fresh fish	[44]
Microorganisms (Safety)		Microbial loads		PLSR	r^2p = 0.97, SEP = 0.38 log CFU/g in flesh side of fillet r^2p = 0.94, SEP = 0.53 log CFU/g in skin side of fillet r^2p = 0.82, SEP = 0.82 log CFU/g in minced samples	
Freshness (Quality)	Atlantic salmon *Salmo salar* (fillets)	Storage time	FT-NIR	PCA	Clear separation between fresh and non-fresh fillets	[45]
Microorganisms (Safety)		Numbers of bacteria		PLSR	r^2cv = 0.64, RMSECV = 0.32 log cfu/g of bacteria	

Application	Species (sample preparation)	Measured property	Instrument	Chemometric method	Statistical results	Reference
Freshness (Quality)	Golden pompano *Trachinotus ovatus* (portions)	Storage time	Vis–NIR	PCA	Samples stored for 0, 2 and 3 days could not be well discriminated	[46]
Freshness (Quality)	Rohu *Labeo rohita* (sliced pieces)	Storage time	Absorption NIR	PCA	Samples were well grouped by their freshness	[47]
Freshness (Quality)	Atlantic cod *Gadus morhua* (fillets)	Storage time	Interactance Vis–NIR	PLSR	$r^2cv = 0.93$, RMSECV = 1.66 days	[48]
Unfrozen vs. Thawed (Authenticity)		Unfrozen vs. Thawed		KNN	> 96% correct classification	
Freshness (Quality)	Atlantic salmon *Salmo salar* L (fillets)	Storage time	Interactance Vis–NIR	PLSR	$r^2cv = 0.95$, RMSECV = 1.91 days	[49]
Unfrozen vs. Thawed (Authenticity)		Unfrozen vs. Thawed		KNN	100% correct classification	
Freshness (Quality)	Cod *Gadus morhua* (fillets)	Quality index method (QIM)	Transflection Vis–NIR	PLSR	$r^2p > 0.93$, RMSEP <3.1 pints	[50]
Wild vs. Farmed (Authenticity)	Seabass *Dicentrarchus labrax* (minced)	Wild vs. Farmed	Reflectance NIR	PLS-DA with measured chemistry, PLS-DA with estimated chemistry, PLS-DA and WPTER	100% correct classification with PLS-DA_mc 94% correct classification with PLS-DA_ec 91% correct classification with PLS-DA_NIR 91% correct classification with WPTER	[53]
Species identification (Authenticity)	Salmon *Salmo salar*, blue whiting *Micromesistius poutassau*, mackerel *Scomber scombrus* and herring *Clupea harengus* (Fishmeal)	Species identification	Reflectance NIR	DPLS and LDA	>80% correct classification with DPLS > 70% correct classification with LDA	[55]

Application	Species (sample preparation)	Measured property	Instrument	Chemometric method	Statistical results	Reference
Species identification (Authenticity)	Winter cod and cod, red mullet and Atlantic mullet, and samlet and salmon trout (skin and meat of fillets)	Species identification	Handheld reflectance NIR	SIMCA	100% correct classification of cod and winter cod 100% correct classification of red mullet and Atlantic mullet 100% correct classification of salmon trout and samlet	[56]
Species identification (Authenticity)	Red mullet *Mullus surmuletus* and plaice *Pleuronectes platessa*, Atlantic mullet	Higher from lower economical value species	FT-NIR	LDA and SIMCA	100% correct classification of red mullet and Atlantic mullet >67% correct classification of plaice and flounder	[57]
Unfrozen vs. Thawed (Authenticity)	*Pseudupeneus prayensis* and flounder *Platichthys flesus flesus* (fillets)	Unfrozen vs. Thawed Atlantic mullet	FT-NIR	LDA and SIMCA	> 82% correct classification of unfrozen and thawed Atlantic mullet	
Species identification (Authenticity)	Silver carp *Hypophthalmichthys molitrix*, black carp *Mylopharyngodon piceus*, bighead carp *Aristichthys nobilis*, grass carp *Ctenopharyngodon idellus*, common carp *Cyprinus carpio*, crucian *Carassius auratus* and bream *Parabramis pekinensis* (minced)	Species identification	Reflectance NIR	PCA-LDA, PLS-LDA, CARS-LDA and FFT - LDA	> 96% correct classification for PCA-LDA > 89% correct classification for PLS-LDA > 88% correct classification for CARS-LDA > 96% correct classification for FFT-LDA	[58]
Species identification (Authenticity)	White croaker *Argyrosomus argentatus*, hairtail *Trichiurus haumela*, and redcoat *Nemipterus virgatus* (surimi)	Species identification	Reflectance NIR	PC-DA	100% correct classification for the validation set	[59]

Application	Species (sample preparation)	Measured property	Instrument	Chemometric method	Statistical results	Reference
Species identification (Authenticity)	Atlantic cod *G. morhua* and haddock *Melanogrammus aeglefinus* (raw fillets and patties)	Species identification	Reflectance handheld NIR and FT-NIR	LDA and SIMCA	> 76% correct classification for portable NIR and LDA 100% correct classification for FT-NIR and LDA > 60% correct classification for portable NIR and SIMCA > 91% correct classification for FT-NIR and SIMCA	[60]
Wild vs. Farmed (Authenticity)	Sea bass *Dicentrarchus labrax* (minced)	Wild or farmed and rearing system	Reflectance NIR	OPLS-DA	100% correct classification for production method > 67% correct classification for farming system	[62]
Geographical origin (Authenticity)		Areas of the Mediterranean Sea		OPLS-DA	> 85% correct classification	
Proximate composition (Quality)	Alaskan pollock *Gadus chalcogrammu*, Atlantic cod *Gadus morhua*, European plaice *Pleuronectes platessa*, common sole *Solea solea*, and turbot *Psetta maxima* (fillets)	Protein, lipids, and moisture	Reflectance NIR	PLSR	$r^2p = 0.80$, RMSEP = 0.15 g/100 g for lipids $r^2p = 0.80$, RMSEP = 0.95% for protein $r^2p = 0.79$, RMSEP = 1.01% for moisture	[64]
Wild vs. Farmed (Authenticity)		Wild vs. farmed		LR and SVM	90% correct classification for sole with LR 99% correct classification for turbot with SVM	
Unfrozen vs. Thawed (Authenticity)		Unfrozen vs. Thawed		XGB, LDA and RF	88% correct classification for cod with XGB 100% correct classification for plaice with LDA 90% correct classification for sole with RF	

Application	Species (sample preparation)	Measured property	Instrument	Chemometric method	Statistical results	Reference
Geographical origin (Authenticity)	Anchovies *Engraulis encrasicholus* (minced semi-finished and finished salted)	Mediterranean country	Reflectance FT-NIR	OPLS-DA	> 95% correct classification	[65]
Unfrozen vs. Thawed (Authenticity)	Red sea bream *Pagrus major* (whole fish)	Unfrozen vs. Thawed	Interactance NIR	LDA and SIMCA	> 63% correct classification for SIMCA > 79% correct classification for LDA	[72]
Unfrozen vs. Thawed (Authenticity)	Swordfish *Xiphias gladius L* (cutlets)	Unfrozen vs. Thawed	Vis–NIR and NIR + monocromator	PLS-DA	> 96.7% correct classification for VIS/NIR > 90.0% correct classification for NIR	[73]
Unfrozen vs. Thawed (Authenticity)	Tuna *Thunnus thynnus* (fillets)	Unfrozen vs. Thawed	Reflectance Vis–NIR	PLS-DA	92% correct classification for unfrozen 82% correct classification for frozen–thawed	[74]
Unfrozen vs. Thawed (Authenticity)	Tilapia (*Oreochromis*) (fillets)	Freeze-thawing cycles	FT-NIR	Mahalanobis distance-DA	> 60.0% correct classification for once frozen–thawed fillets > 60.0% correct classification for samples subjected to several freezing cycles	[75]
Unfrozen vs. Thawed (Authenticity)	Bigeye tuna *Thunnus obesus* (fillets)	Unfrozen vs. Thawed	Reflectance handheld NIR	PLS-DA	91% of correct classification	[76]
Spoilage compounds (Safety)	Silver carp *Hypophthalmichthys molitrix* (sliced portions)	Trimethylamine content	FT-NIR	PLS, Si-PLS and GA-PLS	r^2p = 0.94, RMSEP = 5.75 mg N/100 g with PLS r^2p = 0.95, RMSEP = 5.45 mg N/100 g with Si-PLS r^2p = 0.98, RMSEP = 5.10 mg N/100 g with GA-PLS	[77]

Application	Species (sample preparation)	Measured property	Instrument	Chemometric method	Statistical results	Reference
Spoilage compounds (Safety)	Tuna *Katsuwonus pelamis* and *Thunnus albacares* (minced raw and canned)	Histamine content	Reflectance FT-NIR	OPLSR	$r^2 p > 0.97$, RMSEP <10 mg kg^{-1}	[78]
Spoilage compounds (Safety)	Chub Mackerel *Scomber japonicus* (homogenized and ultrafiltered)	Cadaverine content	NIR	PLSR	$r^2p > 0.98$	[80]
Spoilage compounds (Safety)	Skipjack tuna *Katsuwonus pelmis* (dry extracts)	Histamine content	Reflectance FT-NIR	PLSR	$r^2c > 0.76$, SEP < 3.47 ppm	[81]
Spoilage compounds (Safety)	Silver carp *Hypophthal michthys molitrix* (whole fish, measured in the eye)	K-value	Reflectance FT-NIR	PLS, i-PLS, Si-PLS, SVMR and Si-SVMR	Rp = 93.7%, RMSEP = 0.036 for Si-SVMR	[82]
Spoilage compounds (Safety)	Mackerel oil (extracted from minced samples)	Free fatty acids	Transmittance NIR	MLR and PLS	Rp > 0.862%, SEP < 0.154% for MLR Rp > 0.441, 0.862%, SEP < 0.23, 0.131% for PLS	[83]
Spoilage compounds (Safety)	Bighead carp *Aristichthys nobilis* (minced)	pH, TBARS, total volatile basic nitrogen (TVB-N) and K-value	Reflectance NIR	CARS-PLSR	Rp = 0.945, RMSEP = 0.081 for pH Rp = 0.932, RMSEP = 2.099 for TVB-N Rp = 0.954, RMSEP = 0.107 for TBARS Rp = 0.807, RMSEP = 6.509 for K-value	[85]
Microorganisms (Safety)	Flounder (fillets)	Total bacteria content	Portable NIR	PLSR and genetic algorithm with BP-ANN	$r^2p = 0.916$, RMSEP = 0.40 with PLS $r^2p = 0.966$, RMSEP = 0.083 with BP-ANN	[87]

Author details

Sonia Nieto-Ortega*, Rebeca Lara, Giuseppe Foti, Ángela Melado-Herreros and Idoia Olabarrieta
AZTI, Food Research, Basque Research and Technology Alliance (BRTA), Parque Tecnológico de Bizkaia, Derio, Spain

*Address all correspondence to: snieto@azti.es

References

[1] Siesler HW. Basic principles of near-infrared spectroscopy. In: Burns DA, Ciurczac EW, editors. Handbook of Near-infrared Analysis. 3rd ed. Boca Raton, FL: CRC press; 2008. pp. 7-20

[2] Ozaki Y, Morisawa Y. Principles and characteristics of NIR spectroscopy. In: Ozaki Y, Huck C, Tsuchikawa S, Engelsen SB, editors. Near-Infrared Spectroscopy: Theory, Spectral Analysis, Instrumentation, and Applications. Singapore: Springer Singapore; 2021. pp. 11-35. DOI: 10.1007/978-981-15-8648-4_2

[3] Ozaki Y. Infrared spectroscopy—mid-infrared, near-infrared, and far-infrared/terahertz spectroscopy. Analytical Sciences. 2021;**37**:1193-1212. DOI: 10.2116/analsci.20R008

[4] Beć KB, Grabska J, Huck CW. Near-infrared spectroscopy in bio-applications. Molecules. 2020;**25**:2948. DOI: 10.3390/molecules25122948

[5] El-Azazy M. Introductory Chapter: Infrared spectroscopy - a synopsis of the fundamentals and applications. In: El-Azazy M, editor. Infrared Spectroscopy - Principles, Advances, and Applications. London: IntechOpen; 2018. DOI: 10.5772/intechopen.82210

[6] Brereton RG, Jansen J, Lopes J, Marini F, Pomerantsev A, Rodionova O, et al. Chemometrics in analytical chemistry—part I: History, experimental design and data analysis tools. Analytical and Bioanalytical Chemistry. 2017;**409**: 5891-5899. DOI: 10.1007/s00216-017-0517-1

[7] Brereton RG, Jansen J, Lopes J, Marini F, Pomerantsev A, Rodionova O, et al. Chemometrics in analytical chemistry—part II: Modeling, validation, and applications. Analytical and Bioanalytical Chemistry. 2018;**410**: 6691-6704. DOI: 10.1007/s00216-018-1283-4

[8] Blanco M, Villarroya I. NIR spectroscopy: A rapid-response analytical tool. TrAC Trends in Analytical Chemistry. 2002;**21**:240-250. DOI: 10.1016/S0165-9936(02)00404-1

[9] Rinnan Å, Fvd B, Engelsen SB. Review of the most common pre-processing techniques for near-infrared spectra. TrAC Trends in Analytical Chemistry. 2009;**28**:1201-1222. DOI: 10.1016/j.trac.2009.07.007

[10] de Lima MD, Barbosa R. Methods of authentication of food grown in organic and conventional systems using chemometrics and data mining algorithms: A review. Food Analytical Methods. 2019;**12**:887-901. DOI: 10.1007/s12161-018-01413-3

[11] Yu H, Guo L, Kharbach M, Han W. Multi-way analysis coupled with near-infrared spectroscopy in food industry: Models and applications. Food. 2021;**10**: 802. DOI: 10.3390/foods10040802

[12] Nobari-Moghaddam H, Tamiji Z, Akbari-Lakeh M, Khoshayand MR, Haji-Mahmoodi M. Multivariate analysis of food fraud: A review of NIR based instruments in tandem with chemometrics. Journal of Food Composition and Analysis. 2022;**107**: 104343. DOI: 10.1016/j.jfca.2021.104343

[13] McVey C, McGrath TF, Haughey SA, Elliott CT. A rapid food chain approach for authenticity screening: The development, validation and transferability of a chemometric model

using two handheld near infrared spectroscopy (NIRS) devices. Talanta. 2021;**222**:121533. DOI: 10.1016/j.talanta.2020.121533

[14] Melado-Herreros A, Nieto-Ortega S, Olabarrieta I, Gutiérrez M, Villar A, Zufía J, et al. Postharvest ripeness assessment of 'Hass' avocado based on development of a new ripening index and Vis-NIR spectroscopy. Postharvest Biology and Technology. 2021;**181**: 111683. DOI: 10.1016/j.postharvbio.2021.111683

[15] Ncama K, Magwaza LS, Mditshwa A, Tesfay SZ. Application of visible to near-infrared spectroscopy for non-destructive assessment of quality parameters of fruit. In: El-Azazy M, editor. Infrared Spectroscopy - Principles, Advances, and Applications. London: IntechOpen; 2018. DOI: 10.5772/intechopen.80069

[16] Coppola D, Lauritano C, Palma Esposito F, Riccio G, Rizzo C, de Pascale D. Fish waste: From problem to valuable resource. Marine Drugs. 2021; **19**:116. DOI: 10.3390/md19020116

[17] Ruiz-Salmón I, Laso J, Margallo M, Villanueva-Rey P, Rodríguez E, Quinteiro P, et al. Life cycle assessment of fish and seafood processed products – A review of methodologies and new challenges. Science of The Total Environment. 2021;**761**:144094. DOI: 10.1016/j.scitotenv.2020.144094

[18] Fox M, Mitchell M, Dean M, Elliott C, Campbell K. The seafood supply chain from a fraudulent perspective. Food Security. 2018;**10**: 939-963. DOI: 10.1007/s12571-018-0826-z

[19] Sheng L, Wang L. The microbial safety of fish and fish products: Recent advances in understanding its significance, contamination sources, and control strategies. Comprehensive Reviews in Food Science and Food Safety. 2021;**20**:738-786. DOI: 10.1111/1541-4337.12671

[20] Ye B, Chen J, Fu L, Wang Y. Application of nondestructive evaluation (NDE) technologies throughout cold chain logistics of seafood: Classification, innovations and research trends. LWT-Food Science and Technology. 2022;**158**: 113127. DOI: 10.1016/j.lwt.2022.113127

[21] Ndraha N, Hsiao H-I, Vlajic J, Yang M-F, Lin H-TV. Time-temperature abuse in the food cold chain: Review of issues, challenges, and recommendations. Food Control. 2018;**89**:12-21. DOI: 10.1016/j.foodcont.2018.01.027

[22] Yu W. Applications of near infrared spectroscopy for fish and fish products quality: A review. IOP Conference Series: Earth and Environmental Science. 2021; **657**:012115. DOI: 10.1088/1755-1315/657/1/012115

[23] Shenk JS, Workman JJ, Westerhaus MO. Application of NIR spectroscopy to agricultural products. In: Burns DA, Ciurczac EW, editors. Handbook of Near-infrared Analysis. 3rd ed. Boca Raton, FL: CRC press; 2008. pp. 347-386

[24] Mendez J, Mendoza L, Cruz-Tirado JP, Quevedo R, Siche R. Trends in application of NIR and hyperspectral imaging for food authentication. Scientia Agropecuaria. 2019;**10**:143-161. DOI: 10.17268/sci.agropecu.2018.01.16

[25] Qu J-H, Liu D, Cheng J-H, Sun D-W, Ma J, Pu H, et al. Applications of near-infrared spectroscopy in food safety evaluation and control: A review of recent research advances. Critical Reviews in Food Science and Nutrition.

2015;**55**:1939-1954. DOI: 10.1080/10408398.2013.871693

[26] Power A, Cozzolino D. How fishy is your fish? Authentication, provenance and traceability in fish and seafood by means of vibrational spectroscopy. Applied Sciences. 2020;**10**:4150. DOI: 10.3390/app10124150

[27] Franceschelli L, Berardinelli A, Dabbou S, Ragni L, Tartagni M. Sensing technology for fish freshness and safety: A review. Sensors. 2021;**21**:1373. DOI: 10.3390/s21041373

[28] Liu D, Zeng X-A, Sun D-W. NIR spectroscopy and imaging techniques for evaluation of fish quality—A review. Applied Spectroscopy Reviews. 2013;**48**: 609-628. DOI: 10.1080/05704928.2013.775579

[29] Venugopal V. Biosensors in fish production and quality control. Biosensors and Bioelectronics. 2002;**17**: 147-157. DOI: 10.1016/S0956-5663(01)00180-4

[30] Yeannes MI, Almandos ME. Estimation of fish proximate composition starting from water content. Journal of Food Composition and Analysis. 2003;**16**:81-92. DOI: 10.1016/S0889-1575(02)00168-0

[31] Liu Y, Ma D-H, Wang X-C, Liu L-P, Fan Y-X, Cao J-X. Prediction of chemical composition and geographical origin traceability of Chinese export tilapia fillets products by near infrared reflectance spectroscopy. LWT - Food Science and Technology. 2015;**60**: 1214-1218. DOI: 10.1016/j.lwt.2014.09.009

[32] Nilsen H, Esaiassen M, Heia K, Sigernes F. Visible/near-infrared spectroscopy: A new tool for the evaluation of fish freshness? Journal of Food Science. 2002;**67**:1821-1826. DOI: 10.1111/j.1365-2621.2002.tb08729.x

[33] Karlsdottir MG, Arason S, Kristinsson HG, Sveinsdottir K. The application of near infrared spectroscopy to study lipid characteristics and deterioration of frozen lean fish muscles. Food Chemistry. 2014;**159**:420-427. DOI: 10.1016/j.foodchem.2014.03.050

[34] Isaksson T, Tøgersen G, Iversen A, Hildrum KI. Non-destructive determination of fat, moisture and protein in salmon fillets by use of near-infrared diffuse spectroscopy. Journal of the Science of Food and Agriculture. 1995;**69**:95-100. DOI: 10.1002/jsfa.2740690115

[35] Wold JP, Isaksson T. Non-destructive determination of fat and moisture in whole atlantic salmon by near-infrared diffuse spectroscopy. Journal of Food Science. 1997;**62**: 734-736. DOI: 10.1111/j.1365-2621.1997.tb15446.x

[36] Xiccato G, Trocino A, Tulli F, Tibaldi E. Prediction of chemical composition and origin identification of european sea bass (*Dicentrarchus labrax* L.) by near infrared reflectance spectroscopy (NIRS). Food Chemistry. 2004;**86**:275-281. DOI: 10.1016/j.foodchem.2003.09.026

[37] Trocino A, Xiccato G, Majolini D, Tazzoli M, Bertotto D, Pascoli F, et al. Assessing the quality of organic and conventionally-farmed European sea bass (*Dicentrarchus labrax*). Food Chemistry. 2012;**131**:427-433. DOI: 10.1016/j.foodchem.2011.08.082

[38] Khodabux K, L'Omelette MSS, Jhaumeer-Laulloo S, Ramasami P, Rondeau P. Chemical and near-infrared determination of moisture, fat and

protein in tuna fishes. Food Chemistry. 2007;**102**:669-675. DOI: 10.1016/j.foodchem.2006.05.057

[39] dos Santos DA, Coqueiro A, Gonçalves TR, Carvalho JC, Bezerra JS Jr, Matsushita M, et al. Omega-3 and omega-6 determination in nile Tilapia's fillet based on MicroNIR spectroscopy and multivariate calibration. Journal of the Brazilian Chemical Society. 2020;**31**: 1883-1890. DOI: 10.21577/0103-5053.20200082

[40] Huang Y, Rogers TM, Wenz MA, Cavinato AG, Mayes DM, Bledsoe GE, et al. Detection of sodium chloride in cured salmon roe by SW–NIR spectroscopy. Journal of Agricultural and Food Chemistry. 2001;**49**:4161-4167. DOI: 10.1021/jf001177f

[41] Huang Y, Cavinato AG, Mayes DM, Bledsoe GE, Rasco BA. Nondestructive prediction of moisture and sodium chloride in cold smoked atlantic salmon (*Salmo salar*). Journal of Food Science. 2002;**67**:2543-2547. DOI: 10.1111/j.1365-2621.2002.tb08773.x

[42] Huang Y, Cavinato AG, Mayes DM, Kangas LJ, Bledsoe GE, Rasco BA. Nondestructive determination of moisture and sodium chloride in cured atlantic salmon (*Salmo salar*) (Teijin) using short-wavelength near-infrared spectroscopy (SW-NIR). Journal of Food Science. 2003;**68**:482-486. DOI: 10.1111/j.1365-2621.2003.tb05698.x

[43] Chun H-N, Kim B, Shin H-S. Evaluation of a freshness indicator for quality of fish products during storage. Food Science and Biotechnology. 2014; **23**:1719-1725. DOI: 10.1007/s10068-014-0235-9

[44] Lin M, Mousavi M, Al-Holy M, Cavinato AG, Rasco BA. Rapid near infrared spectroscopic method for the detection of spoilage in rainbow trout (*Oncorhynchus mykiss*) fillet. Journal of Food Science. 2006;**71**:S18-S23. DOI: 10.1111/j.1365-2621.2006.tb12400.x

[45] Tito NB, Rodemann T, Powell SM. Use of near infrared spectroscopy to predict microbial numbers on Atlantic salmon. Food Microbiology. 2012;**32**: 431-436. DOI: 10.1016/j.fm.2012.07.009

[46] Zhang X, Zhou H, Chang L, Lou X, Li J, Hui G, et al. Study of golden pompano (*Trachinotus ovatus*) freshness forecasting method by utilising Vis/NIR spectroscopy combined with electronic nose. International Journal of Food Properties. 2018;**21**:1257-1269. DOI: 10.1080/10942912.2018.1440239

[47] Dei J, Naskar H, Dasgupta S, Nag S, Behera BK, Tudu B, et al. Fish freshness assessment using NIR spectroscopy. In: Proceedings of the 2020 International Conference on Emerging Frontiers in Electrical and Electronic Technologies (ICEFEET); 10-11 July 2020; Patna, India. 2020. pp. 1-3. DOI: 10.1109/ICEFEET49149.2020.9187016

[48] Sivertsen AH, Kimiya T, Heia K. Automatic freshness assessment of cod (*Gadus morhua*) fillets by Vis/Nir spectroscopy. Journal of Food Engineering. 2011;**103**:317-323. DOI: 10.1016/j.jfoodeng.2010.10.030

[49] Kimiya T, Sivertsen AH, Heia K. VIS/NIR spectroscopy for non-destructive freshness assessment of Atlantic salmon (*Salmo salar* L.) fillets. Journal of Food Engineering. 2013;**116**: 758-764. DOI: 10.1016/j.jfoodeng.2013.01.008

[50] Nilsen H, Esaiassen M. Predicting sensory score of cod (*Gadus morhua*) from visible spectroscopy. LWT - Food

Science and Technology. 2005;**38**:95-99. DOI: 10.1016/j.lwt.2004.05.001

[51] European Regulation (EU) n. 1379/ 2013. Available from: https://eur-lex.europa.eu/legal-content/EN/TXT/HTML/?uri=CELEX:32013R1379&from=EN [Accessed: 2022-05-09]

[52] Ghidini S, Varrà MO, Zanardi E. Approaching authenticity issues in fish and seafood products by qualitative spectroscopy and chemometrics. Molecules. 2019;**24**:1812. DOI: 10.3390/molecules24091812

[53] Ottavian M, Facco P, Fasolato L, Novelli E, Mirisola M, Perini M, et al. Use of near-infrared spectroscopy for fast fraud detection in seafood: Application to the authentication of wild european sea bass (*Dicentrarchus labrax*). Journal of Agricultural and Food Chemistry. 2012;**60**:639-648. DOI: 10.1021/jf203385e

[54] Blanco-Fernandez C, Ardura A, Masiá P, Rodriguez N, Voces L, Fernandez-Raigoso M, et al. Fraud in highly appreciated fish detected from DNA in Europe may undermine the Development Goal of sustainable fishing in Africa. Scientific Reports. 2021;**11**: 11423. DOI: 10.1038/s41598-021-91020-w

[55] Cozzolino D, Chree A, Scaife JR, Murray I. Usefulness of near-infrared reflectance (NIR) spectroscopy and chemometrics to discriminate fishmeal batches made with different fish species. Journal of Agricultural and Food Chemistry. 2005;**53**:4459-4463. DOI: 10.1021/jf050303i

[56] O'Brien N, Hulse CA, Pfeifer F, Siesler HW. Near infrared spectroscopic authentication of seafood. Journal of Near Infrared Spectroscopy. 2013;**21**: 299-305. DOI: 10.1255/jnirs.1063

[57] Alamprese C, Casiraghi E. Application of FT-NIR and FT-IR spectroscopy to fish fillet authentication. LWT - Food Science and Technology. 2015;**63**:720-725. DOI: 10.1016/j.lwt.2015.03.021

[58] Lv H, Xu W, You J, Xiong S. Classification of freshwater fish species by linear discriminant analysis based on near infrared reflectance spectroscopy. Journal of Near Infrared Spectroscopy. 2017;**25**:54-62. DOI: 10.1177/0967033516678801

[59] Zhang X-Y, Hu W, Teng J, Peng H-H, Gan J-H, Wang X-C, et al. Rapid recognition of marine fish surimi by one-step discriminant analysis based on near-infrared diffuse reflectance spectroscopy. International Journal of Food Properties. 2017;**20**:2932-2943. DOI: 10.1080/10942912.2016.1261153

[60] Grassi S, Casiraghi E, Alamprese C. Handheld NIR device: A non-targeted approach to assess authenticity of fish fillets and patties. Food Chemistry. 2018; **243**:382-388. DOI: 10.1016/j.foodchem.2017.09.145

[61] Costa C, D'Andrea S, Russo R, Antonucci F, Pallottino F, Menesatti P. Application of non-invasive techniques to differentiate sea bass (*Dicentrarchus labrax*, L. 1758) quality cultured under different conditions. Aquaculture International. 2011;**19**:765-778. DOI: 10.1007/s10499-010-9393-9

[62] Ghidini S, Varrà MO, Dall'Asta C, Badiani A, Ianieri A, Zanardi E. Rapid authentication of European sea bass (*Dicentrarchus labrax* L.) according to production method, farming system, and geographical origin by near infrared spectroscopy coupled with chemometrics. Food Chemistry. 2019;

280:321-327. DOI: 10.1016/j.foodchem.2018.12.075

[63] Vidal NP, Goicoechea E, Manzanos MJ, Guillén MD. Fourier transform infrared spectroscopy as a tool to study farmed and wild sea bass lipid composition. Journal of the Science of Food and Agriculture. 2014;**94**:1340-1348. DOI: 10.1002/jsfa.6417

[64] Gonçalves DB, Santos CSP, Pinho T, Queirós R, Vaz PD, Bloore M, et al. Near infrared reflectance spectroscopy coupled to chemometrics as a cost-effective, rapid, and non-destructive tool for fish fraud control: Monitoring source, condition, and nutritional value of five common whitefish species. Journal of AOAC International. 2020;**104**:53-60. DOI: 10.1093/jaoacint/qsaa114

[65] Varrà MO, Ghidini S, Ianieri A, Zanardi E. Near infrared spectral fingerprinting: A tool against origin-related fraud in the sector of processed anchovies. Food Control. 2021;**123**:107778. DOI: 10.1016/j.foodcont.2020.107778

[66] Hassoun A, Shumilina E, Di Donato F, Foschi M, Simal-Gandara J, Biancolillo A. Emerging techniques for differentiation of fresh and frozen-thawed seafoods: Highlighting the potential of spectroscopic techniques. Molecules. 2020;**25**:4472. DOI: 10.3390/molecules25194472

[67] Hong H, Luo Y, Zhou Z, Bao Y, Lu H, Shen H. Effects of different freezing treatments on the biogenic amine and quality changes of bighead carp (*Aristichthys nobilis*) heads during ice storage. Food Chemistry. 2013;**138**:1476-1482. DOI: 10.1016/j.foodchem.2012.11.031

[68] Leygonie C, Britz TJ, Hoffman LC. Impact of freezing and thawing on the quality of meat: Review. Meat Science. 2012;**91**:93-98. DOI: 10.1016/j.meatsci.2012.01.013

[69] Nakazawa N, Okazaki E. Recent research on factors influencing the quality of frozen seafood. Fisheries Science. 2020;**86**:231-244. DOI: 10.1007/s12562-020-01402-8

[70] Tolstorebrov I, Eikevik TM, Bantle M. Effect of low and ultra-low temperature applications during freezing and frozen storage on quality parameters for fish. International Journal of Refrigeration. 2016;**63**:37-47. DOI: 10.1016/j.ijrefrig.2015.11.003

[71] Bozzetta E, Pezzolato M, Cencetti E, Varello K, Abramo F, Mutinelli F, et al. Histology as a valid and reliable tool to differentiate fresh from frozen-thawed fish. Journal of Food Protection. 2012;**75**:1536-1541. DOI: 10.4315/0362-028x.jfp-12-035

[72] Uddin M, Okazaki E, Turza S, Yumiko Y, Tanaka M, Fukuda Y. Non-destructive visible/NIR spectroscopy for differentiation of fresh and frozen-thawed fish. Journal of Food Science. 2005;**70**:c506-c510. DOI: 10.1111/j.1365-2621.2005.tb11509.x

[73] Fasolato L, Balzan S, Riovanto R, Berzaghi P, Mirisola M, Ferlito JC, et al. Comparison of visible and near-infrared reflectance spectroscopy to authenticate fresh and frozen-thawed swordfish (*Xiphias gladius* L). Journal of Aquatic Food Product Technology. 2012;**21**:493-507. DOI: 10.1080/10498850.2011.615103

[74] Reis MM, Martínez E, Saitua E, Rodríguez R, Pérez I, Olabarrieta I. Non-invasive differentiation between fresh and frozen/thawed tuna fillets using near infrared spectroscopy (Vis-NIRS).

LWT – Food Science and Technology. 2017;**78**:129-137. DOI: 10.1016/j.lwt.2016.12.014

[75] Wang W-L, Chen W-H, Tian H-Y, Liu Y. Detection of frozen-thawed cycles for frozen tilapia (*Oreochromis*) fillets using near infrared spectroscopy. Journal of Aquatic Food Product Technology. 2018;**27**:609-618. DOI: 10.1080/10498850.2018.1461156

[76] Nieto-Ortega S, Melado-Herreros Á, Foti G, Olabarrieta I, Ramilo-Fernández G, Gonzalez Sotelo C, et al. Rapid differentiation of unfrozen and frozen-thawed tuna with non-destructive methods and classification models: Bioelectrical impedance analysis (BIA), near-infrared spectroscopy (NIR) and time domain reflectometry (TDR). Food. 2022;**11**:55. DOI: 10.3390/foods11010055

[77] Agyekum AA, Kutsanedzie FYH, Mintah BK, Annavaram V, Zareef M, Hassan MM, et al. Rapid and nondestructive quantification of trimethylamine by FT-NIR coupled with chemometric techniques. Food Analytical Methods. 2019;**12**:2035-2044. DOI: 10.1007/s12161-019-01537-0

[78] Ghidini S, Chiesa LM, Panseri S, Varrà MO, Ianieri A, Pessina D, et al. Histamine control in raw and processed tuna: A rapid tool based on NIR spectroscopy. Food. 2021;**10**:885. DOI: 10.3390/foods10040885

[79] Liu Y, Wang C, Xia Z, Chen J. Nondestructive evaluation of biogenic amines in crayfish (*Prokaryophyllus clarkii*) by near infrared spectroscopy. Journal of Near Infrared Spectroscopy. 2021;**29**:330-338. DOI: 10.1177/09670335211054298

[80] Shim K, Jeong Y. Freshness evaluation in chub mackerel (*Scomber japonicus*) using near-infrared spectroscopy determination of the cadaverine content. Journal of Food Protection. 2019;**82**:768-774. DOI: 10.4315/0362-028x.jfp-18-529

[81] Pochanagone S, Rittiron R. Preliminary study on the determination of ppm-level concentration of histamine in tuna fish using a dry extract system for infrared coupled with near-infrared spectroscopy. ACS Omega. 2019;**4**:19164-19171. DOI: 10.1021/acsomega.9b02438

[82] Ding R, Huang X, Han F, Dai H, Teye E, Xu F. Rapid and nondestructive evaluation of fish freshness by near infrared reflectance spectroscopy combined with chemometrics analysis. Analytical Methods. 2014;**6**:9675-9683. DOI: 10.1039/c4ay01839g

[83] Zhang H-Z, Lee T-C. Rapid near-infrared spectroscopic method for the determination of free fatty acid in fish and its application in fish quality assessment. Journal of Agricultural Food Chemistry. 1997;**45**:3515-3521. DOI: 10.1021/jf960643r

[84] Yildiz G, Wehling RL, Cuppett SL. Method for determining oxidation of vegetable oils by near-infrared spectroscopy. Journal of the American Oil Chemists' Society. 2001;**78**:495-502. DOI: 10.1007/s11746-001-0292-1

[85] Zhou JJ, Wu XY, Chen Z, You J, Xiong SB. Evaluation of freshness in freshwater fish based on near infrared reflectance spectroscopy and chemometrics. LWT-Food Science and Technology. 2019;**106**:145-150. DOI: 10.1016/j.lwt.2019.01.056

[86] Cheng J-H, Dai Q, Sun D-W, Zeng X-A, Liu D, Pu H-B. Applications of non-destructive spectroscopic techniques for fish quality and safety evaluation and

inspection. Trends in Food Science & Technology. 2013;**34**:18-31. DOI: 10.1016/j.tifs.2013.08.005

[87] Duan C, Chen C, Khan MN, Liu Y, Zhang R, Lin H, et al. Non-destructive determination of the total bacteria in flounder fillet by portable near infrared spectrometer. Food Control. 2014;**42**: 18-22. DOI: 10.1016/j.foodcont. 2014.01.023

Chapter 6

Monitoring Brain Activities Using fNIRS to Avoid Stroke

Yun-Hsuan Chen and Mohamad Sawan

Abstract

Functional near-infrared spectroscopy (fNIRS) is an emerging wearable neuroimaging technique based on monitoring the hemodynamics of brain activity. First, the operation principle of fNIRS is described. This includes introducing the absorption spectra of the targeted molecule: the oxygenated and deoxygenated hemoglobin. Then, the optical path formed by emitters and detectors and the concentration of the molecules is determined using Beer-Lambert law. In the second part, the advantages of applying fNIRS are compared with other neuroimaging techniques, such as computed tomography and magnetic resonance imaging. The compared parameters include time and spatial resolution, immobility, etc. Next, the evolution of the fNIRS devices is shown. It includes the commercially available systems and the others under construction in academia. In the last section, the applications of fNIRS to avoid stroke are presented. The challenges of achieving good signal quality and high user comfort monitoring on stroke patients are discussed. Due to the wearable, user-friendly, and accessibility characteristics of fNIRS, it has the potential to be a complementary technique for real-time bedside monitoring of stroke patients. A stroke risk prediction system can be implemented to avoid stroke by combining the recorded fNIRS signals, routinely monitored physiological parameters, electronic health records, and machine learning models.

Keywords: neuroimaging, functional near-infrared spectroscopy (fNIRS), hemodynamics, monitoring, wearable devices, stroke

1. Introduction

Studying the brain signals helps to gain knowledge of its function. Clinically, cortical signal recordings provide information for researchers to investigate the mechanisms of the cerebral diseases and help the clinicians identify the type of diseases, locate the lesions, and further prescribe the treatments and medications [1, 2]. Therefore, various neuroimaging techniques flourished this century [3]. These techniques are implemented in various applications these decades.

Neuroimaging techniques can be separated into invasive and non-invasive types. The former includes electrocorticography (ECoG), intracortical implants, stereo-electroencephalography (SEEG), spikes, and local field potential (LPF). The latter includes scalp encephalography (EEG), computed tomography (CT), functional magnetic resonance imaging (fMRI), magnetoencephalography (MEG), photoacoustic

IntechOpen

imaging (PAI), functional near-infrared spectroscopy (fNIRS), and Positron emission tomography (PET). Invasive approaches encounter more clinical issues, such as infection, biocompatibility, etc. Non-invasive techniques, such as CT, fMRI, MEG, and PET, can achieve high temporal and spatial resolution. However, these tools are bulky and not easy to access, which is not appropriate for real-time monitoring [4]. Thus, lightweight, compact systems like EEG and fNIRS attract attention [5]. EEG records the electrical signals of many neurons, resulting in a low-spatial resolution. In comparison, fNIRS is an approach to overcome the limitations mentioned above [6].

fNIRS is often used to study brain activity during cognitive and motion tasks [7]. Clinically, fNIRS is applied to diagnose, monitor the progression of the diseases, predict the outcomes of the disorders, and track the effect of rehabilitation [8]. The brain function changes and the evolution of the brain diseases or the recovery process suggest the effectiveness of treatments or rehabilitation. Furthermore, the fNIRS biomarkers can be identified to distinguish the abnormalities and further stratify the severity of the diseases. With the development of various neuromodulation approaches, such as transcranial direct current stimulation (tDCS), transcranial magnetic stimulation (TMS), and ultrasound, combining fNIRS and these techniques to alter the brain activity, a closed-loop brain-machine/–computer interface can be achieved [9, 10]. These closed-loop systems can be used for abstinence of addiction, to reduce the symptoms of epileptic seizures, to release the migraine, and even to enhance attention [11, 12].

While fNIRS have been applied to various neural diseases, this chapter brings an overview of the application to stroke, with special attention to stroke prevention and prediction. Stroke is the second leading cause of death worldwide. Over 50% of stroke patients suffer from various levels of disability after the onset resulting in heavy family and social burdens [13]. However, it is reported in the countries' guidelines that 85% of stroke is preventable. A stroke happens when the blood flow in the brain is interrupted or significantly reduced. This is highly related to the changes in the amount of hemoglobin, which can be detected using fNIRS. This neuroimaging technique has been applied in various stroke cases, such as stroke prediction in people with specific diseases, monitoring during the treatment, stroke management in the perioperative period, and rehabilitation evaluation [14]. Since stroke is known as a chronic disease, investigations are carried out to predict the sign of the stroke onset by tracking the risk factors changes. With user-friendly, compact, and light-weighted characteristics, fNIRS is a promising tool to monitor the hemodynamic parameters continuously or in a high-frequency mode. Combining the real-time monitored physiological parameters and the electronic health records (EHRs), a prediction model can be established by introducing machine learning models [15].

In this chapter, a detailed introduction to the mechanism of neurovascular coupling (Section 2.1), the principles of fNIRS (Section 2.2), the advantages of fNIRS (Section 2.3), the evolutions of fNIRS devices, and the commercially available or self-developed devices (Section 2.4) are presented. Then, the potential of using fNIRS to avoid stroke is elaborated in Section 3. Section 4 is the conclusion.

2. Functional near-infrared spectroscopy (fNIRS)

In this section, an overview of fNIRS, an emerging neuroimaging technique, is presented. The principle, the advantages, and the evolution of the fNIRS systems are discussed in these sections.

2.1 Neurovascular coupling

Neurovascular coupling is a mechanism that the hemoglobin concentration variations result from brain activities. This is the phenomenon that an fNIRS can monitor. When the neurons are in the rest state, the concentrations of the surrounding oxygenated hemoglobin (OxyHb) and the deoxygenated hemoglobin (DeoxyHb) are in a particular state to support the normal metabolism (**Figure 1a**). When the neurons are activated, they need oxygen to support their activation (**Figure 1b**). Therefore, the blood flow of the vessels around the neurons increases, increasing OxyHb to provide additional oxygen to the neurons. Though the neurons consume the oxygen, converting the OxyHb to DeoxyHb during activation, the increased OxyHb is more than the OxyHb which is transformed. Therefore, an increase of OxyHb and a reduction of DeoxyHb are observed during this neuronal activation phase, called neurovascular coupling (**Figure 1c**).

2.2 Principles of fNIRS

The incident lights at the wavelengths that can react the most with the target molecules are applied to monitor the concentration variation optically. The target molecules mean OxyHb and DeoxyHb, whose concentrations are altered due to brain activities. However, OxyHb and DeoxyHb are not the only molecules in the brain. Water is another molecule that accounts for a large percentage of the brain.

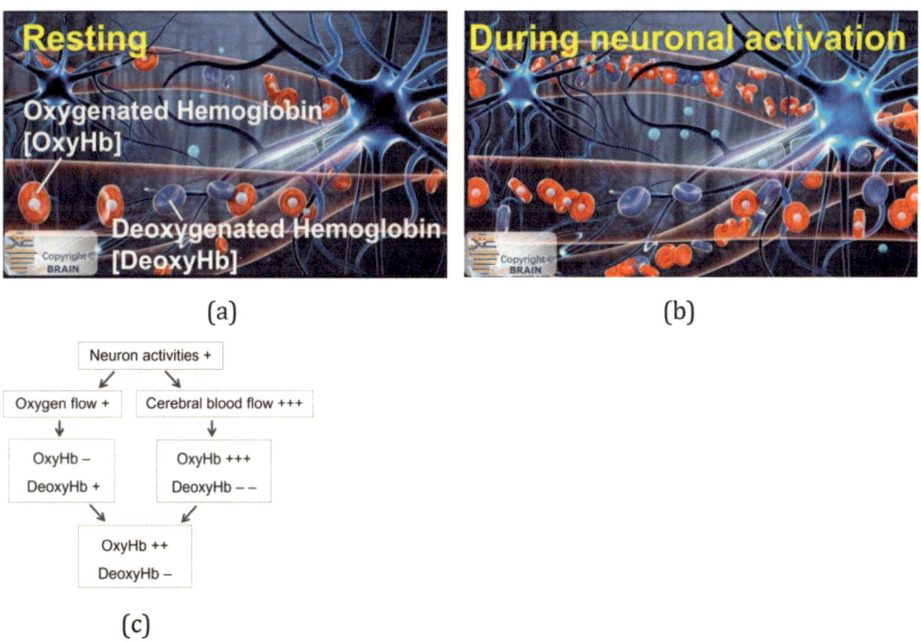

Figure 1.
Schematics of the neurovascular coupling: (a) At rest state, the concentration of oxygenated and deoxygenated hemoglobins (OxyHb and DeoxyHb) is at a certain metabolic state, (b) during neuronal activation, the blood flow, meaning the OxyHb increases to provide the needed oxygen to the activated neurons. While more OxyHb is supplied, more DeoxyHb is generated due to the consumption of the oxygen from the OxyHb molecules, and (c) the overview of the variation of OxyHb and DeoxyHb resulted from the neurovascular coupling when neurons are activated. The red cell is the oxygenated hemoglobin. White circles on the red cells are oxygen. When the oxygen is provided to the neurons, the red cell converts to the blue cell, called deoxygenated hemoglobin.

The absorption spectra of these three molecules are shown in **Figure 2**. An optical window with the wavelength from 700 to 900 nm is an accepted range for the public to detect the concentration of OxyHb and DeoxyHb with limited interferences from the water molecules. To maximize the detection accuracy, fNIRS devices are often equipped with light emitters at 760 nm and 850 nm, which is the wavelength of maximum absorption factors of DeoxyHb and OxyHb, respectively.

After penetrating the cerebral tissues, the incident light is collected by detectors. The detected light is attenuated due to the scattering, absorption, and transmission effects of the interaction of the light and the medium (**Figure 3**). Beer-Lambert law defines the relationship between attenuation and the attenuating species' concentration and the optical path length [6]. For the near-infrared spectroscopy (NIRS) applied on the scalp, since the optical path between the emitter and the detector is not straight, a modified Beer-Lambert law with differential path length factor (DPF) as an additional parameter is applied. Besides, another parameter, G, representing photon loss due to light scattering, is added in the modified formula. Assuming DPF and d are known and remain constant, G remains constant, and the attenuation difference can determine the concentration difference of the attenuating species.

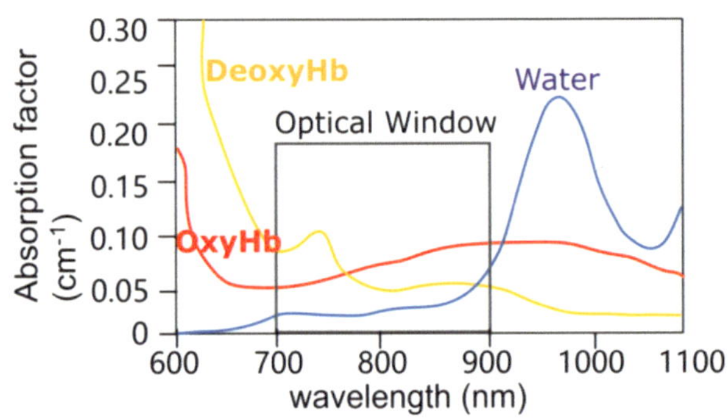

Figure 2.
Absorption spectra of the molecules in the brain. fNIRS is an optical approach to monitor the concentration variation of HbOxy and HbDeoxy.

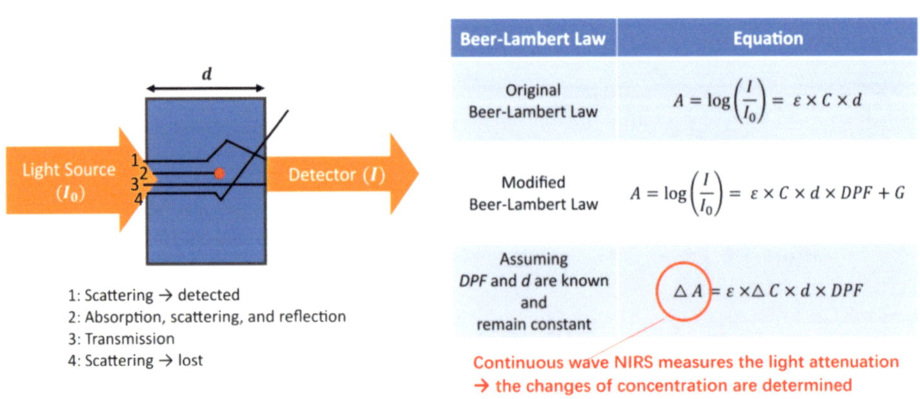

Figure 3.
Modified beer-Lambert law is applied to the recorded fNIRS signals to determine the concentration changes of HbOxy and HbDeoxy: I_0: The intensity of the inserted light; I: The intensity of the detected light; a: Attenuation; ε: Attenuation factor; C: Concentration of attenuating species; d: Optical path length; G: Photon loss due to light scattering, which depends mainly on geometrical factors introduced; DPF: Differential path length factor.

Since it is challenging to define the value G, most of the available NIRS devices can monitor the relative concentration, meaning the variation of concentration, not the absolute value of the concentration of the hemoglobins. This is the imaging principle of a continuous-wave NIRS (cw-NIRS).

To obtain the absolute values of hemoglobins, time-domain and frequency-domain modalities are designed, known as TD-NIRS and FD-NIRS. Diffused optical tomography (DOT) is an upgrade version of TD-NIRS to achieve a high spatial resolution and distinguish the signals from different depth layers under the scalp [16]. This NIRS approach being applied to study the functional changes of the brain cortex during various activities is later known as functional near-infrared spectroscopy, shortened as fNIRS. This shares the same idea as functional MRI (fMRI), to investigate the brain's activated region during certain tasks.

Figure 4 presents an example of the location arrangement of fNIRS's optical electrodes (optodes), meaning the emitters and the detectors. A banana-shaped optical path is formed between each emitter and detector pair. The concentration changes of the hemoglobin passing through the path region can be determined using the modified Beer-Lambert law. The commonly used distance between an emitter and a detector is 3 cm, attaining the deepest sensing depth at 1-1.5 cm beneath the scalp. This shallow layer is mostly located with capillary vessels. Depending on the region of interest, the emitters and detectors can be placed alternately or in other patterns on the scalp. Illustrations in **Figure 4** show the optical path of a cw-fNIRS. If applying a TD-fNIRS, the photons that arrive at different times can be detected, giving the hemodynamic information of various depths of the tissue.

2.3 Comparison of fNIRS and other neuroimaging techniques

Magnetic resonance imaging (MRI), computed tomography (CT), and magnetoencephalography (MEG) are the most common techniques to characterize the abnormalities in the brain (**Figure 5**) [17]. Though MRI and CT offer brain images with a high spatial resolution within the millimeter range, the temporal resolution is low, impeding their application in real-time monitoring. In addition, MRI and CT tests are rather expensive and time-consuming. The MRI, CT, and MEG tests equipment is bulky and should occupy a custom-designed room. Therefore, the patients need to be transported for every measurement. Moreover, the MRI measurements

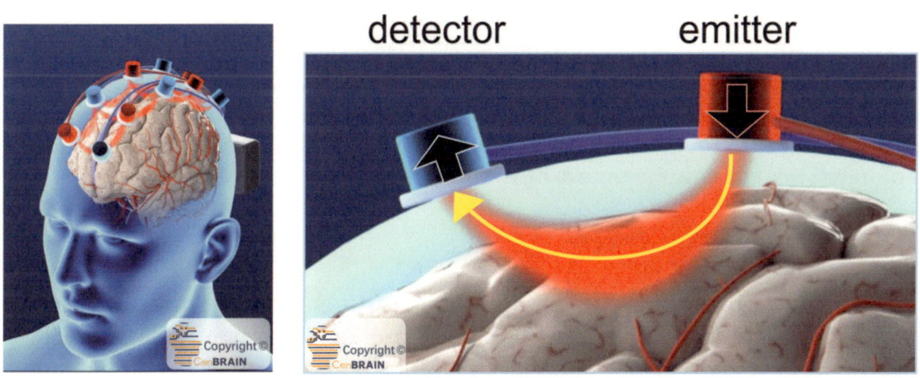

Figure 4.
Placement of fNIRS optodes on the scalp. The hemoglobin changes in the blood vessels located in the optical path formed by a pair of emitter and detector can be monitored.

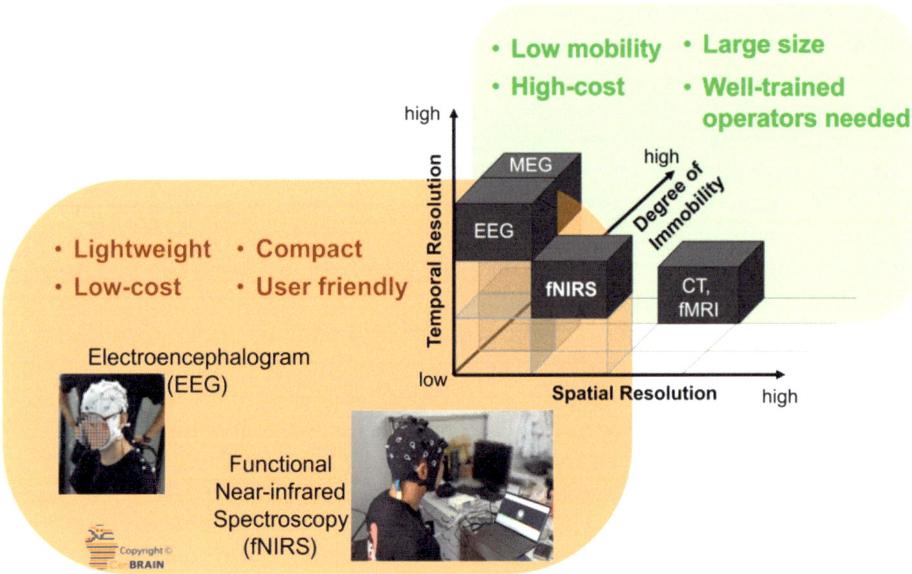

Figure 5.
Comparison of various types of neuroimaging techniques: CT (computed tomography); EEG (electroencephalogram); fMRI (functional magnetic resonance imaging); fNIRS (functional near-infrared spectroscopy); and MEG (magnetoencephalography).

cause loud noise and are sensitive to motions, reducing its usability in a certain group of patients, such as infants. Besides, there are other restrictions for receiving MRI measurements, such as for patients with a pacemaker or metallic implants due to the magnetic fields. CT has the additional disadvantage that patients are exposed to ionizing radiations. Due to the discussed disadvantages, there is no appropriate neuroimaging technique to real-time monitor the hemodynamic response in routine clinical practice.

fNIRS uses pairs of optodes, including a near-infrared light source and a detector, to monitor the concentration change of OxyHb, DeoxyHb, and cerebral blood flow resulting from the brain's hemodynamic activity [18]. Both EEG and fNIRS recording systems are non-invasive, flexible, low-cost, compact, light-weighted, small-sized, portable, undemanding set-up and can be used ambulatory. In addition, they both have less physical restriction and less motion artifact interference compared with functional MRI (fMRI) and CT. Moreover, they have no high intensity (>1 T) magnetic field or ionizing radiations compared with fMRI and CT. Regarding EEG, the spatial resolution is not increased since the recorded signal is generated by the bunch of neurons close to the electrode. The studies using fNIRS to monitor the variations of pathophysiological parameters are introduced in the following paragraphs.

2.4 Evolution of fNIRS devices

The components forming an fNIRS system are the emitter, the detector, and the recording module. The emitters can be a laser, laser diode, or light-emitting diode (LED). The detectors can be photomultiplier tubes (PMTs), silicon photodiodes (SPDs), avalanche photodiodes (APDs), or a novel type called silicon photomultiplier (SiPM). SiPM overcomes the shortcomings of the bulky size of PMTs and the lower sensitivity to light resulting in a limited detection depth of the brain [19].

The evolution of fNIRS technologies in terms of the parameters related to signal quality and user-friendly is shown in **Table 1**. The early fNIRS was often equipped with fiber laser as emitter, therefore using mainline power resulting in a rather bulky system and low mobility. The subjects wearing the cap with optic fiber bundles need to stay around the device and experience movements restriction. A compact fNIRS system with LED powered by batteries is designed to become a wearable device to extend the application scenarios. The fiber laser fNIRS devices can be weighted up to 100 kg, while the LED devices' weight is around 500 g for the helmet, including the optodes.

The oximeter is the original device using the NIRS principle to detect the concentration variation of the hemoglobin [20]. The device is a clip to be placed on the finger or two patches attached to the forehead to detect the oxygen saturation of the blood (SpO_2). These oximeters are single sites with emitters of often two wavelengths each, which are used to detect the OxyHb and DeoxyHb. The oximeters are considered simple prototypes of various emerging increasing complexity fNIRS devices. Increasing the number of optode channels is mandatory to simultaneously monitor the hemodynamic states of multiple areas of the cerebral cortex. The challenges of upgrading a single site to multiple sites fNIRS are applying the frequency encoding method to minimize the interferences from other optical paths. After overcoming this issue, the channel number of fNIRS boomed. It is noted that one single light emitter can create multiple channels if numerous detectors surround it. Using laser fibers, around one hundred channels can be reached. However, achieving these numerous channels requires heavy cables and emitters mounted on the cap. The weighty system causes discomfort. In addition, the relatively rigid cables limit the number of optodes can be placed on the scalp. In contrast, the channel number of fNIRS easily achieves one to two thousand channels if using LED as emitters. Thus, the hemodynamic conditions of the brain cortex's large region of interest can be imaged.

Due to the limitation of the light penetration, cw-fNIRS is only sensitive to the hemodynamic variation of the vessels at 1-2 cm beneath the scalp, around half of the commonly used distance between the light source and the detector, 3 cm.

Parameters	Evolution Trends	
	From	To
Size	Bulky	Light and compact
Mobility	Fixed location and motion restricted due to mainline powered	Wireless and wearable thanks to batteries operated
Channel Number	Single channel	Optical fibers: ~100 channels LED: can reach 1000-2000 channels
Penetration Depth	cw-fNIRS: 1-2 cm	TD-fNIRS: various depths can be monitored
Imaging reconstruction ability	Single site	High spatial resolution using high-density diffuse optical tomography (HD-DOT)
Flexibility for multimodal neuroimaging	fNIRS itself only	fNIRS-EEG, fNIRS-PET, fNIRS-fMRI, etc.

Table 1.
Evolution of the key parameters of the fNIRS devices.

With TD-fNIRS, the OxyHb and DeoxyHb concentration changes at various depths of the cerebral tissue can be determined [21].

Only the hemodynamic states of light scattered spots can be determined with limited fNIRS channels. High density diffused optical tomography (HD-DOS) is designed to construct the 3D images of hemodynamics based on NIRS principles [16]. This HD-DOS contained various distances between the emitter and the detector, from 1 to 6 cm, to increase the spatial resolutions of a conventional cw-fNIRS.

As introduced in Section 2.3, every neuroimaging technique has its advantages and shortcomings. Therefore, applying multiple them simultaneously compensates for the shortcomings of each other and achieves a more comprehensive view of the brain activity. The fNIRS-EEG system can simultaneously record the neuron electrical signals and hemodynamic signals, which helps investigate the neurovascular coupling effect. However, fNIRS-fMRI obtains cross-validation for brain activity. The high-temporal-resolution fNIRS and the high-spatial-resolution fMRI are complementary.

Concerning the commercially available fNIRS devices, around 20 companies worldwide, such as NIRx, Artinis, ISS, SHIMADZU, g.tec, etc. Most of their products are cw-fNIRS; only one company, ISS Inc., introduced the multi-channel FD-fNIRS. Also, no company offers a TD-fNIRS yet. Both laser fiber and LED are used in these commercially available systems for the light emitters.

Regarding the emerging platforms in academia, research groups are focusing on gaining more flexibility when applying fNIRS in specific applications. Most of them focus on miniaturization and improving the temporal and spatial resolutions. A review paper summarized fNIRS hardware implanted in academia from 2012 to 2017 [22]. In recent 5 years, a 16-source and 16-detector fNIRS intended for epilepsy study was recently published [23]. An open-source, light-weighted (142 g) fNIRS is developed [24]. WearLight, a wearable fNIRS, is designed [25]. Growing research has been conducted to implement TD-fNIRS to compensate for the missing product in the market [26].

Furthermore, many multimodal fNIRS-EEG systems are designed to better explore both the electrical and hemodynamic signals of the neurovascular coupling [27]. The co-author of this chapter is one of the first few group to present the multimodal fNIRS-EEG system [28]. This prototype was further optimized and proved to be valid when monitoring signals from patients with epilepsy and stroke [29]. In addition, Mobile Modular Multimodal Biosignal Acquisition (M3BA architecture) is an architecture that can wirelessly transmit EEG, fNIRS, and accelerometer signals [30].

3. Applications of fNIRS on stroke

The stroke-related applications using fNIRS include predictions, diagnosis, monitoring during treatments, outcome prediction, stroke management, and rehabilitations [14]. fNIRS monitoring during the treatments is the most challenging part since the protocol needs to be fine-tuned to reduce the impede on the planned surgeries and treatments. In Section 3.1, the studies that reported the fNIRS measurements during the treatments of stroke patients are summarized. In Section 3.2, the role of fNIRS on stroke risk prediction and prevention is discussed. The challenges of applying fNIRS to avoid stroke are discussed in Section 3.3.

3.1 fNIRS monitoring during treatments

A stroke happens when the blood flow in the brain is interrupted or significantly reduced. Lack of oxygen and nutrients supply to the brain cells results in cell death. Ischemic stroke contributes to 85% of strokes occurs when the blood flow in cerebral vessels is either blocked by a blood clot or reduced by atherosclerosis. Removing the obstruction and restoring the blood flow of the affected area as early as possible reduces the impact of ischemia and improves the outcome of stroke.

Nowadays, intravenous thrombolysis therapy using tissue plasminogen activator (tPA) is the only medicine treatment approved by U.S. Food and Drug Administration for acute ischemic stroke [31]. tPA is an enzyme involved in clot breakdown. However, less than 50% of the patients receiving tPA therapy are successfully recanalized. At least 30% of them encounter the main complication, hemorrhage of tPA. Half of them show no response to the tPA therapy due to the complex conditions of the combination of characteristics of the blood clot and their health conditions [32].

Since another treatment named mechanical thrombectomy (MT) has a longer time window, a higher recanalization rate, fewer contraindications, and fewer complications than intravenous thrombolysis, it has become a more popular approach for treatment. MT improves functional outcome and prevents severe disability and mortality for those eligible for this treatment nowadays [33]. MT is a surgery to remove a blood clot from a blood vessel of patients with large vessel occlusion (LVO). Though the surgical approaches and devices of MT have been rapidly improved these years, 4–29% of the patients encounter complications after receiving MT [34]. Around 6% of patients have hemorrhage and 2-20% have reocclusion. Without immediate detection and proper treatments for these complications, the recanalization of MT treatment is in vain.

Real-time monitoring the hemodynamic status prior to, during, and after the treatments can help detect the abnormalities in time and alert the clinicians to take action. Continuous fNIRS monitoring of the cerebral hemodynamic variations of patients who underwent tPA therapy, known as thrombolysis, was reported in a few publications [35–37]. One research group suggested that the OxyHb and the total hemoglobin concentration (THC) of both ipsilesional and contralesional hemispheres increased during the first 2.5 h after tPA treatment [35]. While another group studied the relative change in regional O_2 saturation (rSO_2) and the interhemispheric rSO2 ($IH\Delta rSO_2$) difference of 1 hour, 6 hours, 12 hours, 18 hours, and 24 hours after the tPA treatment are analyzed and compared with the National Institutes of Health Stroke Scale (NIHSS scores) or either MRI and CT images at the same time frame [36]. A protocol is proposed in a clinical trial to monitor the fNIRS data along with the blood pressure, heart rate, respiratory rate, rSO_2, Glasgow coma scale and the NIHSS at every 15 minutes from the beginning of the thrombolysis till 120 minutes after the starting point [37].

On the other hand, recent and important studies on continuous fNIRS monitoring of the cerebral hemodynamic variations of patients receiving MT treatment are summarized in **Table 2**. Most studies start fNIRS monitoring prior to MT treatment. However, since the preparation of fNIRS monitoring can impede an urgent surgery, the number of measurements starting prior to the MT is limited.

Regarding the locations of optodes, the optodes are placed on the forehead of patients in most studies. While the optodes are placed on the region of interest (ROI), the area is close to the location of ischemic stroke and the opposite hemisphere in only one study. The measured parameters: OxyHb, DeoxyHb, HbT

Studies	Time Frame (prior to, during or after MT)	Number of patients	Number of fNIRS optodes and of the channels	Locations of optodes	Parameters used to access hemodynamic activity
Giacalone et al. [38]	Two times: 1st within 24 h of stroke symptoms onset and 2nd from the first time point up to 5 days	47 patients +35 controls	TD-fNIRS: 12 pairs of optodes	F3-F5, C3-C1, P3-P5 for the left hemisphere and F4-F6, C4-C2, P4-P6 for the right hemisphere. More on ROI.	DeoxyHb, OxyHb, HbT, and StO_2
Hiramatsu et al. [39]	Prior to and during	3	2 light sources and 1 detector -> 1 channel on each side of the forehead	1 on each side of the forehead	rSO_2
Ritzenthaler et al. [40]	Prior, during, and continuous for 24 h after MT	17	2 light sources and 1 detector -> 1 channel on each side of the forehead	1 on each side of the forehead	rSO_2, interhemispheric difference (IHD)
Hametner et al. [41]	Prior[†], during, and continue until patients were extubated or until 6 hours after the intervention	63, 43 are good for analyzing	2 light sources and 1 detector -> 1 channel on each side of the forehead	Upper outer forehead	unilateral rSO_2, IH△rSO_2 average successive rSO_2 variability
Iversen [42]	(1) Prior, during and 2 h after MT (2) 24 h after MT (3) 3 m after MT	100 (target)	8 light sources and 2 detectors	Frontal cortex	CBF, Cerebral autoregulation (CA)
Forti [43]	Prior to, during and after MT	2	TD-DOS and DCS modules: 2 optodes → 1 channel	Forehead	Absolute values of OxyHb & DeoxyHb (via TD-DOS), CBF (via DCS)
Akiyama [44]	Prior to, during and after the CAS	17	2 optodes → 1 channel	Region of interests	OxyHb, DeoxyHb, HbT, Time to peak concentration (TTP)

rSO_2: regional cerebral tissue oxygenation; StO_2: tissue oxygen saturation; ROI: region of interest; CAS: Carotid artery stenting; IHΔrSO_2: interhemispheric rSO_2 difference; DCS: Diffuse correlation spectroscopy; DOS: diffuse optical spectroscopy.
[†]: only 25% of the analyzed data includes recording before the MT.

Table 2.
Studies regarding fNIRS monitoring of acute stroke patients with mechanical thrombectomy recanalization or stenting.

(the combination of the previous two parameters), and rSO_2 (regional cerebral tissue oxygenation) are used to evaluate the efficacy of MT and monitor the hemodynamic states after the surgery. In some studies, the difference between the parameters of two hemispheres over time is characterized. It is found that the variations of the recorded parameters suggest the success of recanalization and are associated with the clinical outcome.

From the above examples, it is concluded that NIRS technique can provide information of real-time changes in hemodynamic response revealing the effects of thrombolysis and thrombectomy. Further applying this information to avoid stroke will be presented in Section 3.2.

3.2 Using fNIRS-EEG to predict and further avoid stroke

Since multimodal fNIRS-EEG is becoming wearable devices with high accessibility for brain monitoring, this section will focus on combining these two techniques. The concept of applying multimodal fNIRS-EEG neuroimaging for stroke risk prediction is shown in **Figure 6**. The first stage is to identify the fNIRS-EEG biomarkers associated with the hemodynamic conditions of stroke onset zones. Studies are launched to monitor the physiological signals of a stroke patient before the treatments are carried out. The measurements continue along with the treatments to the recovering phase in the perioperative period. The variation of the biomarkers when the treatments succeed, the MT or the tPA, or when any complications occur is valuable to train a model for stroke risk prediction. Combining fNIRS and EEG data recordings with other physiological parameters and electronic health records (EHRs) to implement into a machine learning model, a prediction system can be obtained to predict either the onset of complications in stroke patients or on people with a high risk of stroke.

The first milestone to achieve the aim of stroke prediction is real-time tracking of the occurrence of complications after MT or tPA according to the parameters recorded from the multimodal fNIRS-EEG system and the routine clinical recordings.

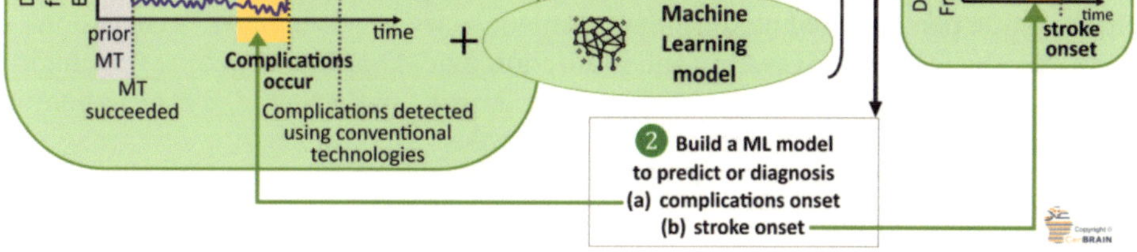

Figure 6.
Combining the fNIRS-EEG, other physiological parameters, EHRs, and ML model, a system can be established to predict the onset of the complications after stroke treatments or the onset of stroke cases in people with a high risk of stroke. DAR: The ratio of the power of the delta band to the power of alpha bands in EEG signals. ML: Machine learning. MT: Mechanical thrombectomy. rSO_2: Regional cerebral tissue oxygenation.

Investigating the mechanisms of the variation of the recorded parameters while combining other available physiological parameters over time distinguish the different impacts on brain activity caused by different complications. When understanding the variation trend of the recorded physiological parameters prior to and during the occurrence of complications after treatments (the yellow time range in **Figure 6**), these parameters are potential biomarkers as an alert for close attention and even further personalized medical or surgical treatments.

The conventional neuroimaging technologies and routine monitored physiological parameters have limitations for stroke management. CT or MRI examinations are impracticable for continuous bedside monitoring of post-treatment patients. Routine measured physiological parameters (i.e., blood pressure (BP), peripheral pulse oximetry (SpO_2), electrocardiography (ECG), heart rate (HR), etc.) indirectly suggest the reduction of blood flow in a certain area of the brain. Promisingly, the fNIRS-EEG recordings provide additional parameters directly reflect the variation of hemodynamic activity and neuron electrical in real-time. The first step to creating the prediction system is to demonstrate this multimodal fNIRS-EEG system is a promising tool to fill in the vacancy of current technologies for post-treatment management.

The second step is to develop a machine learning (ML) model for stroke risk prediction based on the database collected during the monitoring. With the emerging of applying ML models on electronic health records (EHRs) for risk prediction of diseases, the recorded data will be cleaned and organized as a database for future research use. Besides the parameters derived from fNIRS-EEG recordings, the routine measurements of physiological parameters (i.e., BP, SpO_2, ECG, HR, etc.) for post-treatment are included in the database. In addition, the medical history of the recruited patients is included in the database since these factors impact the outcome. Furthermore, the clinical outcomes (the disability and mortality of certain periods after MT) are included. To deal with such a high dimension EHRs and the recorded physiological data, machine learning is an emerging technique to explore the relationship between input features and outcome. Therefore, it is becoming a popular technique for disease prediction. Examples are implementing ML algorithms for predicting seizures using intracortical EEG (iEEG) recordings and predicting bladder volume using a smart neuroprosthesis [45]. The association of the recorded fNIRS-EEG parameters and other data in EHRs with the reduction of blood flow when complications occur is analyzed using machine learning techniques. The importance of features (recorded parameters and patients' health information) is ranked according to the occurrence and progression of reocclusion or hemorrhage during modeling.

With the non-invasive and wearable fNIRS-EEG system and the development prediction model, the variation of selected features recorded from post-treatment can identify or predict the occurrence of complications. In addition, the model can be used to guide personalized medical management based on each patient's condition. Moreover, the fNIRS-EEG system and prediction model applied to people with high stroke risk can suggest the progress of atherosclerosis (accounts for 20% of ischemic stroke) or predict the occurrence of ischemic stroke.

3.3 Challenges of using fNIRS to avoid stroke

The challenges of acquiring high-quality fNIRS signals of brain activity and the corresponding possible solutions have been well explored [14]. The difficulties of applying fNIRS to stroke patients to generate a stroke risk prediction system are introduced here.

3.3.1. Appropriate patients with electrodes and optodes located in the region of interest benefit the most from the fNIRS-EEG measurement

To record the effective signals revealing the occurrence of complications after the treatments, placing the optodes and electrodes on the area that may present the most variation of neuron electrical signals and hemodynamics is a key scientific issue. This issue can be resolved by selecting appropriate patients and relevant locations of the electrodes and optodes.

fNIRS has limited spatial resolution due to the properties of light penetration, thus, the recorded data is sensitive to the hemodynamic changes of the blood vessels located no deeper than 1-2 cm beneath the scalp [46, 47]. Therefore, patients with revascularized vessels located at a shallow layer of the cerebral cortex are appropriate for the proposed EEG-fNIRS measurement. Accordingly, potential complications that occur close to the revascularized vessels can be monitored by the fNIRS-EEG system. Based on this criterion, the patients recruited in the study discussed in **Figure 6** were selected by physicians according to the CT, MRI, or other imaging techniques carried out immediately after admission and after the treatment. CT can distinguish ischemic stroke and hemorrhage stroke. Moreover, the location of infarction and the successfulness of recanalization before and after treatment can be determined. On the other hand, MRI detects the location of ischemia and the affected area with higher precision than CT images.

When potential patients for the measurement are selected, the locations of the fNIRS optodes and EEG electrodes are determined and arranged alternately to cover the area of interest. Such area of interest includes the brain area around the infarcted site and the surgical intervention as well as the symmetric area of the contralesional hemisphere.

3.3.2. Duration of monitoring

Various types of complications occur at different times after treatment. For example, around 6% of patients receiving MT have hemorrhage and 2-20% of them have reocclusion. Hemorrhage often happens within 24 hours of the start of MT. Various types of hemorrhage are related to surgical procedures or the health condition of patients (blood pressure, location of ischemia region, severity of the stroke, etc.). Reocclusion happens when more platelets accumulate at a certain location of a vessel where the surgery takes place due to the damage of the wall of the vessel. In addition, patients with more embolic fragments and stenosis at the thrombectomy site have a high risk of reocclusion. The reocclusion often occurs 24 to 48 hours after MT and is associated with poor outcomes [48].

To monitor the progress of various types of complications, the ideal duration of monitoring should last 24 hours continuously after MT and then at least every 12 hours from 24 hours after MT till the patient is discharged from the hospital.

3.3.3. Data analysis to extract the effective parameters recorded using fNIRS-EEG

Investigating various approaches to handle the recorded data during data analysis to find the optimal parameters suggesting the occurrence and progress of complications is the 3rd key issue. The analyzed data include directly measured signals, derived values, or the combination of the data from fNIRS and EEG recordings.

For fNIRS, the investigated parameters are the change of total hemoglobin concentration (THC) and regional cerebral oxygen saturation (rSO_2) can be computed based on ΔOxyHb and ΔDeoxyHb. The hemodynamics across the hemispheres can be compared by the interhemispheric rSO2 (IHΔrSO2) difference. The association of the recorded parameters with physiological conditions is analyzed to better understand the causes of the variation of parameters.

Regarding EEG, since it is not easy for people unfamiliar with EEG signals to understand the related neurological activity with the corresponding recording, one common approach to analyzing these signals is to convert them to a frequency domain. The converted data inspecting the signal strength as a function of frequency is defined as power spectrum density (PSD). This is useful since it is known that signal patterns of certain frequencies will occur during certain brain activities. It has been studied that the ratio of the power of the delta band to the power of alpha bands (DAR) correlated the most to the infarct area of ischemic stroke [49]. Also, the interhemispheric asymmetry can be characterized using the Brain Symmetry Index (BSI) derived based on the power of signals from the bilateral hemispheres.

The parameters for analysis are not restricted to those discussed above; other possible combinations of the recorded data are also explored. The factors integrated with the direct and indirect parameters from fNIRS and EEG recording can be potential features reflecting the occurrence of complications after stroke treatments.

Besides data collected from the recording facilities, other data recorded through conventional clinical procedures after MT are the key elements to be compared with. Ideally, continuous CT or MRI scans show the most detailed information on the occurrence and progress of complications. However, it is not feasible to attain the scans frequently. Therefore, the vital signs and the rating scales evaluating the impairment caused by a stroke are the parameters to suggest the occurrence of complications or the outcome of recanalization. Routine measurements recommended by worldwide guidelines for post-thrombectomy patients are BP, SpO_2, ECG, heart rate, and blood glucose.

The National Institute of Health Stroke Scale (NIHSS) is the most used score scale for stroke management to quantify neurological dysfunction or deficit after stroke. According to the guidelines regarding the management after thrombectomy, the vital signs and NIHSS are monitored every 30 minutes within the first 6 hours, then every 1 hour from 6 to 24 hours, and then once per 6 hours from 24 to 72 hours. The correlation between the data from fNIRS-EEG recordings and conventional features for detecting complications can suggest the occurrence of complications or stroke onset.

4. Conclusions

We introduced in this chapter the emerging optical technique, fNIRS, to monitor the hemodynamic condition reflecting the activities of the brain. fNIRS consists of emitting light with wavelengths near-infrared range to detect the concentration variation of OxyHb and DeoxyHb. Consequently, the detected light is converted to the absorber's concentration using the modified Beer-Lambert law. fNIRS is an outstanding technique to perform real-time monitoring compared with other common neuroimaging techniques in size, weight, and accessibility. fNIRS has been applied on stroke patients to monitor their hemodynamic conditions in prior-, peri-, and post-operative periods. The variations of images provided by fNIRS, before the onset of complications on stroke patients who received either thrombolysis or thrombectomy treatments are crucial to stroke prediction. Integrating the recorded fNIRS-EEG features, other physiological parameters, and EHRs, the ML model can provide a risk of the stroke prediction system. With the feasibility of continuous or high-frequency fNIRS recordings, the prediction precision of the prediction system to avoid stroke can be increased.

Acknowledgements

The authors disclosed receipt of the following financial support for the research, authorship, and/or publication of this article: This work was supported by Westlake University, grant number [041030080118], and Zhejiang Key R&D Program [2021C03002].

Conflict of interest

The authors declare no conflict of interest.

Notes

The copyrights of the figures in this book chapter belong to CenBRAIN Neurotech, School of Engineering, Westlake University. Please properly cite the origin of the figures if wanting to reuse them in other places.

Author details

Yun-Hsuan Chen[1,2*] and Mohamad Sawan[1,2*]

1 CenBRAIN Neurotech, School of Engineering, Westlake University, Hangzhou, China

2 Institute of Advanced Technology, Westlake Institute for Advanced Study, Hangzhou, China

*Address all correspondence to: chenyunxuan@westlake.edu.cn
and sawan@westlake.edu.cn

References

[1] Zhou Y. Neuroimaging - neurobiology, multimodal and network applications [Internet]. London: IntechOpen; 2020. 140 p. [cited 2022 Jun 06]. Available from: https://www.intechopen.com/books/9347 DOI: 10.5772/intechopen.83577

[2] Zhang Y-D, Dong Z, Wang S-H, Yu X, Yao X, Zhou Q, et al. Advances in multimodal data fusion in neuroimaging: Overview, challenges, and novel orientation. Information Fusion. 2020;**64**:149-187

[3] Rybář M, Daly I. Neural decoding of semantic concepts: A systematic literature review. Journal of Neural Engineering. 2022;**19**(2):021002

[4] Kakkar P, Kakkar T, Patankar T, Saha S. Current approaches and advances in the imaging of stroke. Disease Models & Mechanisms. 2021;**14**(12):dmm048785

[5] Jiaming C, Theodore JH, Pulkit G, Jana MK. Enhanced spatiotemporal resolution imaging of neuronal activity using joint electroencephalography and diffuse optical tomography. Neurophotonics. 2021;**8**(1):1-21

[6] Leon-Carrion J, Leon-Dominguez U. Functional near-infrared spectroscopy (fNIRS): Principles and neuroscientific applications. In: Bright P, editor. Neuroimaging - Methods. London, UK: InTech; 2012

[7] Pinti P, Tachtsidis I, Hamilton A, Hirsch J, Aichelburg C, Gilbert S, et al. The present and future use of functional near-infrared spectroscopy (fNIRS) for cognitive neuroscience. Annals of the New York Academy of Sciences. 2020;**1464**(1):5-29

[8] Rahman MA, Siddik AB, Ghosh TK, Khanam F, Ahmad M. A narrative review on clinical applications of fNIRS. Journal of Digital Imaging. 2020;**33**(5):1167-1184

[9] Larrivee D, Rayegani SM. Neurostimulation and neuromodulation in contemporary therapeutic practice [Internet]. London: IntechOpen; 2020. p. 274 [cited 2022 Jun 06]. Available from: https://www.intechopen.com/books/8059 DOI: 10.5772/intechopen.77890

[10] Paulmurugan K, Vijayaragavan V, Ghosh S, Padmanabhan P, Gulyás B. Brain–computer interfacing using functional near-infrared spectroscopy (fNIRS). Biosensors. 2021;**11**:389. DOI: 10.3390/bios11100389

[11] Curtin A, Tong S, Sun J, Wang J, Onaral B, Ayaz H. A systematic review of integrated functional near-infrared spectroscopy (fNIRS) and transcranial magnetic stimulation (TMS) studies. Frontiers in Neuroscience. 2019;**13**:84. DOI: 10.3389/fnins.2019.00084

[12] Figeys M, Zeeman M, Kim ES. Effects of transcranial direct current stimulation (tDCS) on cognitive performance and cerebral oxygen hemodynamics: A systematic review. Frontiers in Human Neuroscience. 2021;**15**:623315. DOI: 10.3389/fnhum.2021.623315

[13] Beltrán-Rodríguez I, Casado LG-T, Durán-Borrella Ó, González-Feito P, Sánchez-Serrano I, de la Fuente Blanco R. Post-stroke rehabilitation: A necessary step. In: Sanchetee P, editor. Post-Stroke Rehabilitation. London: IntechOpen; 2022

[14] Chen Y-H, Sawan M. Trends and challenges of wearable multimodal technologies for stroke risk prediction. Sensors. 2021;**21**(2):460. DOI: 10.3390/s21020460

[15] Guo Y. A new paradigm of "real-time" stroke risk prediction and integrated Care Management in the Digital Health era: Innovations using machine learning and artificial intelligence approaches. Thrombosis and Haemostasis. 2022;**122**(1):5-7

[16] Wheelock MD, Culver JP, Eggebrecht AT. High-density diffuse optical tomography for imaging human brain function. Review of Scientific Instruments. 2019;**90**(5):051101

[17] Moradi S, Ferdinando H, Zienkiewicz A, Särestöniemi M, Myllylä T. Measurement of cerebral circulation in human. In: Scerrati A, Ricciardi L, Dones F, editors. Cerebral Circulation - Updates on Models, Diagnostics and Treatments of Related Diseases. London: IntechOpen; 2022

[18] Almajidy RK, Mankodiya K, Abtahi M, Hofmann UG. A newcomer's guide to functional near infrared spectroscopy experiments. IEEE Reviews in Biomedical Engineering. 2020;**13**:292-308. DOI: 10.1109/RBME.2019.2944351

[19] Pagano R, Libertino S, Sanfilippo D, Fallica G, Lombardo S. Improvement of sensitivity in continuous wave near infra-red spectroscopy systems by using silicon photomultipliers. Biomedical Optics Express. 2016;**7**(4):1183-1192

[20] Ferrari M, Quaresima V. A brief review on the history of human functional near-infrared spectroscopy (fNIRS) development and fields of application. NeuroImage. 2012;**63**(2):921-935

[21] Torricelli A, Contini D, Pifferi A, Caffini M, Re R, Zucchelli L, et al. Time domain functional NIRS imaging for human brain mapping. NeuroImage. 2014;**85**:28-50

[22] Hubin Z, Robert JC. Review of recent progress toward a fiberless, whole-scalp diffuse optical tomography system. Neurophotonics. 2017;**5**(1):1-11

[23] Liu G, Cui W, Hu X, Xiao R, Zhang S, Cai J, et al. Development of a miniaturized and modular probe for fNIRS instrument. Lasers in Medical Science. 2022;**37**(4):2269-2277

[24] Tsow F, Kumar A, Hosseini SMH, Bowden A. A low-cost, wearable, do-it-yourself functional near-infrared spectroscopy (DIY-fNIRS) headband. HardwareX. 2021;**10**:e00204

[25] Saikia MJ. A Wearable, Configurable Functional Near-Infrared Spectroscopy System for Wireless Neuroimaging. Kingston, USA: University of Rhode Island; 2019

[26] Lacerenza M, Buttafava M, Renna M, Torricelli A, Tosi A, Mora AD, et al., editors. Compact time-domain NIRS oximeter for non-invasive brain and muscle monitoring. In: Applied Industrial Optics 2021. Washington, DC: Optica Publishing Group; 2021

[27] Uchitel J, Vidal-Rosas EE, Cooper RJ, Zhao H. Wearable, integrated EEG-fNIRS technologies: A review. Sensors (Basel). 12 Sep 2021;**21**(18):6106. DOI: 10.3390/s21186106. PMID: 34577313; PMCID: PMC8469799

[28] Sawan M, Salam MT, Lan JL, Kassab A, Gelinas S, Vannasing P, et al. Wireless recording systems: From noninvasive EEG-NIRS to invasive EEG devices. IEEE Transactions on Biomedical Circuits and Systems. 2013;**7**:186-195

[29] Kassab A, Le Lan J, Tremblay J, Vannasing P, Dehbozorgi M, Pouliot P, et al. Multichannel wearable fNIRS-EEG system for long-term clinical monitoring. Human Brain Mapping. 2018;**39**(1):7-23

[30] OpenNIRS.org – Modular open hardware for Near InfraRed Spectroscopy. 2022. Available from: http://opennirs.org/m3ba-mobile-modular-multimodal-biosignal-acquisition/

[31] Marshall RS. Progress in intravenous thrombolytic therapy for acute stroke. JAMA Neurology. 2015;**72**(8):928-934

[32] Çetiner M. Thrombolysis in acute stroke. In: Ambrosi PB, editor. Cerebrovascular Diseases - Elucidating Key Principles. London: IntechOpen; 2021

[33] Giles JA, Vellimana AK, Adeoye OM. Endovascular treatment of acute stroke. Current Neurology and Neuroscience Reports. 2022;**22**(1):83-91

[34] Pilgram-Pastor SM, Piechowiak EI, Dobrocky T, Kaesmacher J, Den Hollander J, Gralla J, et al. Stroke thrombectomy complication management. Journal of NeuroInterventional Surgery. 2021;**13**(10):912

[35] Delgado-Mederos R, Gregori-Pla C, Zirak P, Blanco I, Dinia L, Marín R, et al. Transcranial diffuse optical assessment of the microvascular reperfusion after thrombolysis for acute ischemic stroke. Biomedical Optics Express. 2018;**9**(3):1262-1271

[36] Annus A, Nagy A, Vecsei L, Klivenyi P. 24-hour near-infrared spectroscopy monitoring of acute Ischaemic stroke patients undergoing thrombolysis or Thrombectomy: A pilot study. Journal of Stroke and Cerebrovascular Diseases. 2019;**28**(8):2337-2342

[37] Sanci E. Evaluation of the Effectiveness of Thrombolytic Therapy in Acute Ischemic Stroke Patients Using NIRS (E-NESE). ClinicalTrials.gov; 2018 Available from: https://clinicaltrials.gov/ct2/show/record/NCT03621917?term=NCT03621917&draw=2&rank=1

[38] Giacalone G, Zanoletti M, Re R, Germinario B, Contini D, Spinelli L, et al. Time-domain near-infrared spectroscopy in acute ischemic stroke patients. Neurophotonics. 2019;**6**(1):015003

[39] Hiramatsu R, Furuse M, Yagi R, Ohnishi H, Ikeda N, Nonoguchi N, et al. Limit of intraoperative near-infrared spectroscopy monitoring during endovascular thrombectomy in acute ischemic stroke. Interventional Neuroradiology. 2018;**24**(1):57-63

[40] Ritzenthaler T, Cho TH, Mechtouff L, Ong E, Turjman F, Robinson P, et al. Cerebral near-infrared spectroscopy: A potential approach for Thrombectomy monitoring. Stroke. 2017;**48**(12):3390-3392

[41] Hametner C, Stanarcevic P, Stampfl S, Rohde S, Veltkamp R, Bösel J. Noninvasive cerebral oximetry during endovascular therapy for acute ischemic stroke: An observational study. Journal of Cerebral Blood Flow & Metabolism. 2015;**35**(11):1722-1728

[42] Iversen HK. Monitoring Stroke Patients with Near-Infrared Spectroscopy Before, during and after Endovascular Treatment. ClinicalTrials.gov; 2018. Available from: https://clinicaltrials.gov/ct2/show/NCT03738644

[43] Forti RM, Favilla CG, Cochran JM, Baker WB, Detre JA, Kasner SE, et al. Transcranial optical monitoring of cerebral hemodynamics in acute stroke patients during mechanical Thrombectomy. Journal of Stroke and Cerebrovascular Diseases. 2019;**28**(6):1483-1494

[44] Akiyama T, Ohira T, Hiraga K, Tanizaki Y, Akaji K, Kawase T.

Hemodynamic response following neuronal activity in cases with severe stenosis of the cervical internal carotid artery and its serial change through carotid artery stenting. Journal of Neuroendovascular Therapy. 2008;**2**(2):93-100

[45] Rong G, Mendez A, Bou Assi E, Zhao B, Sawan M. Artificial intelligence in healthcare: Review and prediction case studies. Engineering. 2020;**6**(3):291-301

[46] Quaresima V, Ferrari M. A mini-review on functional near-infrared spectroscopy (fNIRS): Where do we stand, and where should we go? Photonics. 2019;**6**(3):87

[47] Almajidy RK, Mankodiya K, Abtahi M, Hofmann UG. A Newcomer's guide to functional near infrared spectroscopy experiments. IEEE Reviews in Biomedical Engineering. 2020;**13**:292-308

[48] Mosimann Pascal J, Kaesmacher J, Gautschi D, Bellwald S, Panos L, Piechowiak E, et al. Predictors of unexpected early Reocclusion after successful mechanical Thrombectomy in acute ischemic stroke patients. Stroke. 2018;**49**(11):2643-2651

[49] Shreve L, Kaur A, Vo C, Wu J, Cassidy JM, Nguyen A, et al. Electroencephalography measures are useful for identifying large acute ischemic stroke in the emergency department. Journal of Stroke and Cerebrovascular Diseases. 2019;**28**(8):2280-2286

Chapter 7

Application of Infrared Techniques for Characterisation of Vector-Borne Disease Vectors

Maggy Sikulu-Lord and Rafael Maciel-de-Freitas

Abstract

The interest of the entomological community in the application of the Near Infrared (NIR) and Mid Infrared (MIR) spectroscopy techniques for characterisation of insect species has seen a dramatic increase over the last decade. In this chapter, we discuss the application of the Near infrared spectroscopy technique for characterising insects including mosquitoes, triatomine bugs, Culicoides and houseflies under varying environmental and experimental conditions. We focus our discussion on the recent progress made in the application of the NIRS technique to predict the age, species and infection status of mosquito vectors of malaria and arbovirus infections relative to traditional tools such as dissections and molecular techniques and how the tool could be applied in vector surveillance programs to determine disease hotspots and direct current and future interventions. Lastly, we briefly discuss the application of MIR spectroscopy technique for age grading, species identification, blood meal identification and as a potential diagnostic tool for malaria parasites.

Keywords: surveillance, malaria, arbovirus, vector-borne disease, dengue, disease transmission, Zika, chikungunya, age-grading, taxonomic identification

1. Introduction

Entomological and epidemiological surveillance can be broadly described as the continuous monitoring of the transmission potential of vectors and human cases in a given area, respectively. It is an information-based activity involving the collection, analysis, interpretation and appropriate presentation of large volumes of data originating from a variety of sources to monitor the presence of pathogens in a population in a timely manner [1, 2]. Teams with a strong background in entomology, epidemiology, demography, climatology, cartography, statistics and modelling contribute to this exercise creating a multidisciplinary surveillance system to help understand spatiotemporal features of disease transmission [3, 4]. Since one of the most relevant characteristics of surveillance is to inform where and when the risk of disease transmission is higher, it can trigger targeted application of appropriate interventions. Therefore, timely surveillance could help mitigate disease transmission in endemic settings [5, 6].

IntechOpen

Surveillance of insects of medical and veterinary importance is a crucial undertaking required to assess the effectiveness of interventions targeting those insects. For example, prevention of mosquito-borne diseases including malaria and arboviruses such as Zika, dengue, chikungunya and yellow fever, rely on vector control interventions such as insecticide-treated bed nets, indoor residual spraying or the use of repellents as protective measures against infective bites [7]. The effectiveness of these interventions can be determined by assessing changes in insect's behaviour. For example, mosquitoes can change their biting and resting behaviour to avoid contact with insecticide-treated surfaces [8]. Moreover, a change in species composition might mean interventions that normally work against an existing species are no longer effective against the new species [9]. This means routine surveillance to monitor such changes among insect vector populations is a crucial determinant of the effectiveness of interventions.

Estimating survival of insect populations such as mosquitoes to determine how long they have lived is one way of assessing the effectiveness of an intervention. Another method for assessing the effectiveness of interventions against insect populations involves determining the species composition of the target population. This is traditionally achieved through amplification of DNA of the target insect using molecular techniques such as Polymerase Chain Reaction [10]. Effectiveness of interventions can also be assessed by determining whether the target insect population is infected with pathogens such as parasites or viruses. The life cycle of insects that transmit pathogens involves 1) adult emergence from pupae, 2) mating, 3) feeding including blood and sugar feeding, 4) laying eggs and repeating steps 3–4 until the insect eventually dies. When the insect feeds on an infected host, the pathogens infect the midgut of the disease vector where they undergo a period of development known as the extrinsic incubation period (EIP). The EIP usually ranges from 4 to 18 days depending on the insect and the environmental factors such as temperature [11, 12]. Mature pathogens migrate to the salivary glands of the insect and can be inoculated into a host during the next cycle of feeding. PCR [13, 14], enzyme-linked immunosorbent assay [15–17] or dissection coupled with microscopy [18] are the conventional ways of differentiating infected from uninfected insects. The effectiveness of more recent interventions where mosquitoes infected with an intracellular endosymbiont bacteria known as *Wolbachia* [19, 20] to block pathogen transmission [21, 22], are assessed by the number of insects that have maternally acquired the bacteria by PCR or quantitative PCR [23].

Traditional age grading, species identification and infection detection techniques can be time-consuming and costly for large-scale surveillance. These limitations have led to the development of alternative techniques including the application of the near-infrared (NIR) and the Mid-infrared (NIR) spectroscopy. The next sections of the book chapter will discuss the recent applications of the two techniques as alternative techniques for age grading, species identification, infection detection and blood meal analysis of mosquitoes and other insects of medical and veterinary importance.

2. Application of NIRS for characterisation of insects

2.1 Age grading

Age grading involves the assessment of insect vectors to determine how long they have survived. For example, malaria and arbovirus transmitting mosquitoes are only infectious to the host if they are at least 10–12 days old post emergence due to the period required by the pathogen to develop within the mosquito. Estimating age of mosquitoes is therefore crucial as it determines the efficacy of interventions that target survival of those insects. Traditionally, the age of mosquitoes is determined by dissecting their ovaries to determine whether they have previously laid eggs [24] or how many times they have laid eggs [25, 26]. Those that have laid eggs are presumed to successfully have been blood-fed on a host therefore they are considered older and potentially infectious compared to those without an egg-laying history which is normally regarded as young or harmless mosquitoes. Here we will discuss how infrared has been used as an alternative age grading technique for malaria and arbovirus transmitting mosquitoes as well as for fruitflies, houseflies and culicoides.

2.1.1 Houseflies

Houseflies are mechanical vectors of more than 100 diseases caused by viruses, bacteria, parasites and fungus to humans and animals. NIR was first applied to predict the age of houseflies in 2000 [27]. The technique was compared against pteridine fluorescence technique. NIR correctly classified houseflies with a predictive accuracy ranging from 89 to 91% as either 1 day or 10-day-old flies depending on whether the head or the whole body was scanned and whether they were scanned fresh or preserved in ethanol for 2 days. Wavelengths responsible for differentiating young from old flies were identified as 850–950, 1125–1225, 1350–1450, and 1650–1700 nm. These wavelengths correspond to the first, second, and third CH absorption overtones of CH3, CH2, and CH groups and the first combinatioof n CH absorption (**Figure 1**).

2.1.2 Mosquitoes

Mosquitoes were first age graded using NIR in 2009 by Mayagaya and colleagues using ASD NIR spectrometers (ASD, Inc., Boulder, Colorado), [28]. The authors showed that NIR could predict the age of the main African malaria vector *Anopheles gambiae* and *Anopheles arabiensis* reared in the lab with an accuracy of 90% into < 7 or ≥ 7 days age groups. NIR was also used to predict the age of male mosquitoes in the same study. Since then, NIR has been used to predict the age of the two African

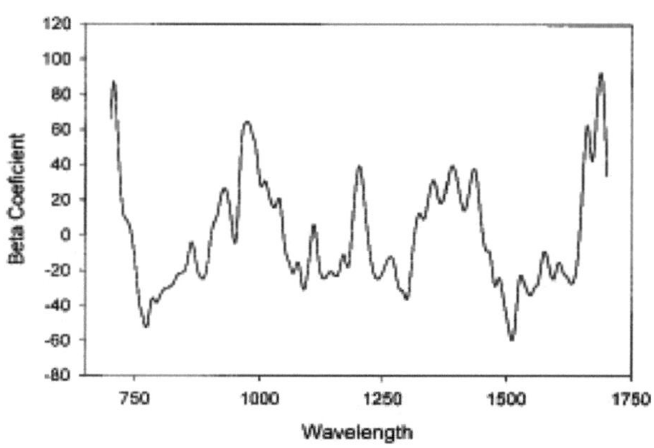

Figure 1.
Regression coefficient graph for age grading houseflies using NIR showing absorption bands in the 1st, 2nd, 3rd overtone regions and combination region [27].

malaria mosquitoes from varying environmental conditions including those reared in a semi-field system SFS [29], those exposed to pyrethroids [30], and more recently to predict the age of the primary vector of arboviruses *Aedes aegypti* [31] and the secondary vector *Aedes albopictus* [32] in the lab with similar predictive accuracy as the malaria vector. Across all species, NIR accurately differentiated very young (1–4 days) old mosquitoes from very old >15 days old mosquitoes but overlaps were observed for middle age groups. This could be due to minimal age-related changes occurring among the middle-aged (5–10 days old) mosquitoes. Recent advances in machine learning techniques have reduced this overlap to produce age prediction models with on a finer scale [33, 34]. NIR age prediction seems to correlate with a number of changes occurring within a mosquito as it undergoes the ageing process such as changes in cuticular hydrocarbons [35], changes in protein expression levels [36, 37] and changes in transcriptional profiles [38] all of which have previously been linked to the ageing process of a mosquito. **Figure 2** shows regression coefficients used to predict the age of female *An. gambiae* [28] and female *Ae. aegypti* [31] mosquitoes with the most informative peaks seen from 1100 nm to 1900 nm for both species.

2.1.3 Culicoides

The ability of NIR QualitySpec Pro spectrometer (ASD, Inc., Boulder) to predict the age of female *Culicoides sonorensis* the blood-feeding species of Ceratopogonidae commonly referred to as biting midges, a vector of bluetongue and other arboviruses was first demonstrated by Reeves and colleagues in 2010 [39]. Three colonies namely VanRyan (VR), Ausma (AU) and AK were used. Using PLS to train models, NIR differentiated younger midges that were 1–3 days old from older midges that were 9–12 days with an overlap observed between adjacent age groups. Younger age groups (1–3 days old) midges were generally predicted as older and older midges (12 days old) were generally predicted as younger (**Figure 3**).

2.1.4 Fruit flies; drosophila

Aw and colleagues have demonstrated the capacity of NIR spectroscopy technique (ASD, Inc., Boulder Colarado) to predict the age of male and female *Drosophila melanogaster* and *Drosophila simulans* using laboratory-reared colonies [40]. Training sets developed from *D. melanogaster* Alst and *D. simulans* Hw strains with seven age groups

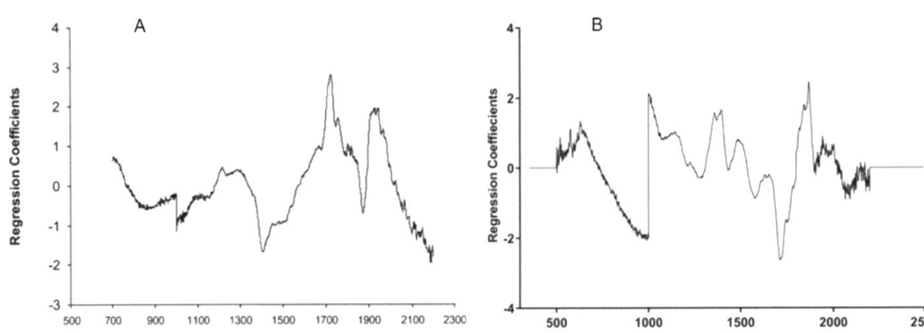

Figure 2.
Regression coefficients for predicting (A) the age of An. gambiae (Ifakara strain) and (B) Ae. aegypti (Australia strain).

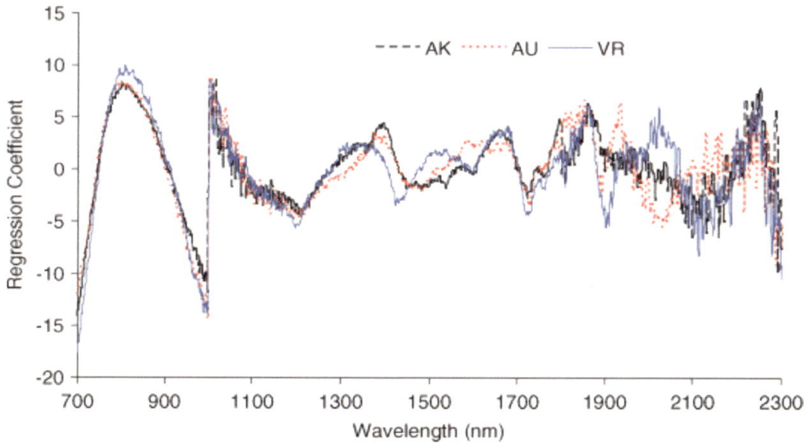

Figure 3.
Regression coefficients used to predict the age of thye three colonies of Culicoides sonorensis [39].

(1,5, 9, 13, 17, 21 and 25 days old) were used to predict independent sets of *D. melanogaster* Dah and *D. simulans* Ky strains at three age groups (5, 13 and 21 days old) into three age groups (<9, 9–18 and > 18 days old). Generally, females were predicted with higher accuracy than males. The prediction accuracy of females that were 5, 13 and 21 days old was 88,86 and 56%, respectively for *D. simulans* Ky and 62, 83 and 84% for *D. melanogaster* Dah whereas the accuracy for predicting the age of males that were 5, 13 and 21 days old was 78, 82 and 37%, respectively for *D. simulans* Ky and 62, 83 and 84% for *D. melanogaster* Dah. Like mosquitoes, female flies were generally predicted more accurately into two groups < 9 (72–88% accuracy) and > 9 days old (88–98% accuracy) than into three groups most probably due to the increasing age-associated changes as the flies got older.

2.2 Species identification

Species identification is crucial for vector control programs as it determines whether existing interventions are effective against all or selected vector species in the target area. A change in mosquito feeding behaviour that is from indoors to

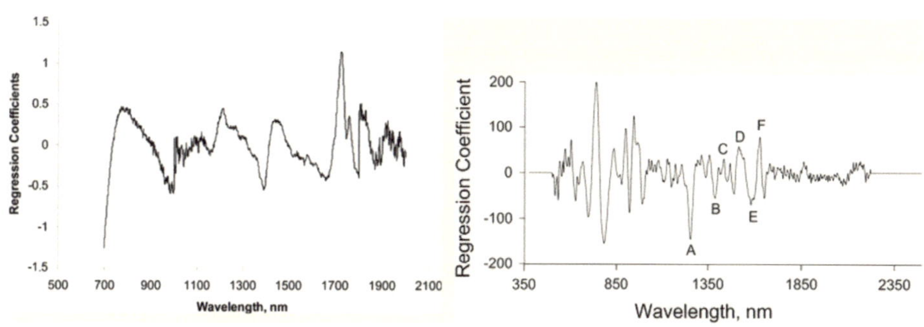

Figure 4.
Regression coefficients for determining the species of (A) female An. gambiae s.s. and An. arabiensis mosquitoes when using 10 partial least squares regression factors [28], and (B) D. melanogaster and D. simulans with important peaks shown with letters A-E [40].

outdoors can lead to significant impacts on the efficacy of existing interventions. That means interventions targeting indoor biting mosquitoes such as bed nets or indoor residual sprays would no longer be effective [9]. Secondly, different species may require different control interventions depending on their breeding, biting or resting behaviour. We discuss here the application of NIR for differentiating cryptic species of mosquitoes and Drosophila.

NIR has been used to differentiate two insect species that would otherwise rely on molecular techniques for differentiation. The major African malaria mosquito *An. gambiae* and *An. arabiensis* are morphologically identical species. NIR has been used to differentiate these two species from laboratory colonies, semi field colonies and from the wild with predictive accuracies of 90% [28, 29] achieved for both laboratory and field collected samples. NIR has also been used to differentiate Drosophila species, *D melanogaster* and *D simulans* with 94% and 82% accuracy, respectively (**Figure 4**) [40].

2.3 Infection detection

One of the fields in which Infrared techniques have recently been successfully applied is the detection of infection in intact insects [41]. This is particularly important in detecting disease hotspots during an outbreak. For infection detection, the NIRS technique has been shown to be simple, cost-effective and rapid relative to molecular techniques such as PCR. In recently published work, NIRS was reported to be 18 times faster and 110 cheaper than RT-qPCR for Zika detection in *Ae. aegypti* mosquitoes [42]. Therefore, facilitating timely entomo-virological surveillance to reduce outbreaks in human population [43, 44].

2.3.1 NIRS detection of arboviruses in mosquitoes

The term arbovirus is an acronym for arthropod-borne virus, that is, virus transmitted through the bite of infected vectors. Arboviruses transmitted to humans by mosquitoes belonging to the family Togaviridae for example, Chikungunya and Ross River viruses or to family Flaviviridae for example, Dengue, Zika, Yellow Fever, Japanese Encephalitis, St. Louis, West Nile viruses etc. [45–47]. With the exception of the African Swine Fever Virus, arboviruses have ribonucleic acid (RNA) in their genome with a positive sense single-stranded RNA (+ssRNA), or a negative sense single-stranded RNA (-ssRNA) [48].

For arboviruses to be transmitted to humans, they require a period of development with an insect. The mosquito cycle begins when a mosquito ingests blood from an infected host during the viremia period and ends with the inoculation of the virus, into a susceptible host. However, an extrinsic incubation period (EIP) is required, whereby an intense viral replication in different organs and tissues of the insect occurs, including the reproductive system, possibly infecting unborn offspring [49]. Depending on the virus type and environmental temperature, viruses and parasites take 5–14 days from the ingestion of an infectious blood meal to reach the salivary glands of mosquitoes [50].

Traditionally, the diagnosis of arbovirus infection in mosquitoes is based on the use of molecular-based techniques such as polymerase chain reaction (PCR), and quantitative PCR (qPCR) [51–54]. However, these techniques are not suitable for large-scale surveillance due to the time, cost and expertise required to process samples. Thus, the development of new techniques to improve surveillance using a fast, reliable and affordable identification of pathogens in disease vectors is an urgent need [41, 55–57].

The development of spectroscopy techniques to detect arbovirus infection has been a growing field of research in the last 5 years. The majority of published results demonstrate the application of NIRS for the detection of arboviruses such as dengue, Zika and chikungunya in intact laboratory-reared *Aedes aegypti* mosquitões [41, 42, 58]. The first attempt to detect arboviruses in mosquito vectors using NIRS was published in 2018 by Fernandes and colleagues who showed that Zika virus (ZIKV) in laboratory-reared *Ae. aegypti* mosquitoes could be detected by simply shining infrared light on the head/thorax or the abdomen of female mosquitoes and using machine learning techniques to differentiate infected from uninfected samples. ZIKV was originally an understudied virus restricted to certain regions of Africa. However, around 2010, it started to spread to other parts of the world, and it was declared a public health emergency of international concern in 2015 following its association with microcephaly in newborns and Guillain-Barré syndrome in adults in South America.

Using Partial Least Square regression models, independent samples of mosquitoes that were either infected with Zika or without Zika infection were predicted with 97.3% and 88.8% when head/thoraces were scanned and 88.8% when abdomens were scanned. The difference in accuracy between heads/thoraces and abdomen is most likely related to virus multiplication in the mosquito body. The highest accuracy of 99.6%, (N = 228) was observed when mosquito heads/thoraces were scanned 7 days post-infection (dpi), which could correlate with the duration of ZIKV EIP in *Ae. aegypti* mosquitoes [42]. Slightly lower prediction accuracy for mosquitoes scanned at 4 and 10 dpi indicates low viraemia in heads/thoraces following this incubation period.

Following successful demonstration that NIR can predict Zika in intact mosquitoes in laboratory-controlled assays, its accuracy in a more natural setting was tested. This involved infecting mosquitoes with ZIKV or chikungunya virus (CHIKV) and scanning the samples 14 dpi, that is, after the virus had sufficient time to complete its EIP within the mosquito. Mosquitoes were killed by aggressively shaking their cage to mimic natural death in the field whilest being in traps. The intention was to avoid the use of any chemical compound that could affect NIR spectra. Traditionally, surveillance systems for arboviruses vectors rely on checking mosquito adult traps once every week. Following the death of mosquitoes, they were individually added to a catching bag of a BG-Sentinel trap, a harsh microenvironment produced by the constantly blowing fan intended to retain trapped mosquitoes. Spectra were individually collected daily over a period of seven days. The findings of this study show that NIR can predict both ZIKV and CHIKV 7 days post their death with predictive accuracy >90% for both arboviruses (**Figure 5**) [58].

2.3.2 NIRS detection of Wolbachia in mosquitoes and drosophila

Wolbachia is a gram-negative intracellular symbiont bacterium that is found naturally infecting arthropods, such as insects, arachnids, crustaceans, and isopods [59–61]. In addition, they are also found in nematodes. This bacterium is present in 40–60% of species, including some mosquitoes such as *Culex quinquefasciatus*, *Astacus fluviatilis* and *Ae. albopictus* [62]. However, the main vectors of malaria and arboviruses; *Anopheles* and *Ae. aegypti* are not naturally infected by this bacterium.

The NIRS technique has been used to detect *Wolbachia* in *D. melanogaster* and *D. simulans*, both of which are natural hosts of *Wolbachia*. The average predictive accuracy was 87% and 92% for *D. melanogaster* and *D. simulans*, respectively [40]. Authors attributed their findings to spectral signatures related to either the presence and the concentration of lipopolysaccharide molecules or the physiological changes caused by

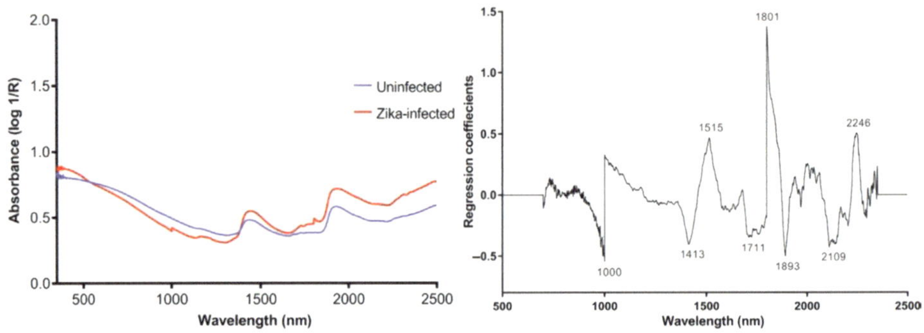

Figure 5.
Average raw spectra (left panel) and regression coefficients of Zika infected and uninfected mosquitoes. Fig adapted from [42].

this bacterium on *Wolbachia*-infected flies. *Wolbachia* is presented in high densities in insect bodies, which can be detected in germ line cells and non-reproductive host tissues [40, 63, 64]. For mosquitoes, NIRS has only been used to detect two species of *Wolbachia* in male and female *Ae. aegypti* [65]. NIRS differentiated females and males infected with *w*MelPop from uninfected ones with an overall accuracy of 96% and 87.5%, respectively whereas *w*Mel infected females and males were predicted with na accuracy of 92% and 89%, respectively. The higher accuracy in detecting *w*MelPop-infection in *Ae. aegypti* females might be due to their existence in higher densities following its transinfection in *Ae. aegypti* compared to *w*Mel. In addition, NIRS differentiated females and males infected with *w*Mel from those infected with *w*MelPop with an accuracy of 96.6% and 84.5%, respectively [65] and in a separate study to detect *w*Mel in mosquitoes 7 days post death (**Figure 6**) [58].

2.3.3 NIRS detection of plasmodium parasites in anopheles mosquitoes

Malaria is an infectious, febrile, potentially serious disease caused by protozoan parasites belonging to the genus Plasmodium transmitted to humans via bites of female infected Anopheles mosquitoes. A total of five species of Plasmodium can be transmitted to humans: *Plasmodium falciparum* is responsible for > 90% of reported annual cases and deaths, mostly in sub-Saharan Africa.

Malaria occurrence is widespread but has a higher prevalence in tropical and subtropical areas. In 2020, 240 million cases of malaria were estimated, resulting in an estimated 627,000 deaths with 95% of the cases and deaths in Sub-saharan Africa.

Esperanca and colleagues demonstrated for the first time that NIR could detect *Plasmodium berghei in laboratory-reared Anopheles stephensi* [66]. The average classification accuracy for differentiating infected from uninfected mosquitoes was 72%. However, NIRS was incapable of determining sporozoite intensity [66]. A more recent publication indicated the technique was capable of detecting oocyst and sporozoite-stage infections with 88% and 95% accuracy, respectively but the prediction accuracy was dependent on parasite concentration [67].

The latest manuscript reported findings on the investigation of the ability of NIRS to identify *P. falciparum* in *An. coluzzii* mosquitoes. Models were developed laboratory-reared infected mosquitoes by using blood from infected subjects. Overall, NIRS correctly classified 73% of mosquitoes as infected or uninfected and differentiated between uninfected and those with either oocysts or sporozoites with a 70% accuracy.

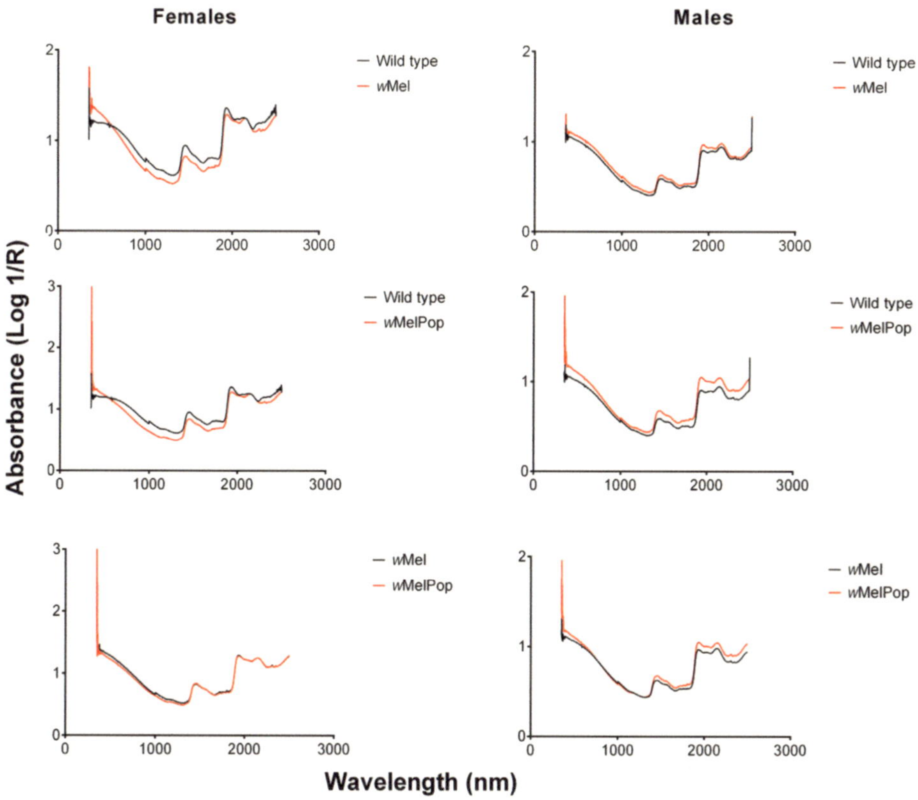

Figure 6.
Shows comparative average raw spectra of wild, wMel and wMelPop Wolbachia strains for female and male Ae. aegypti mosquitoes [65].

However, those models were unable to predict infection in wild-caught mosquitoes [68]. This indicates that robust NIRS models for predicting infection in wild mosquitoes need to be developed from naturally infected mosquitoes.

2.3.4 NIRS detection of Trypanosoma cruzi in Triatomine sp.

Chagas disease (CD), or American trypanosomiasis, is an infectious disease that plagues the entire American continent, from the southern United States to Argentina and Chile. The WHO estimates that the number of infected individuals in Latin America is approximately 6 million people, with about 70 million living at risk of contracting the disease [69]. In recent decades, due to increase mobility between distant regions of the globe, CD has become a global health priority. The disease has become a problem in areas such as Europe and Japan, as a consequence of the immigration of infected individuals acting as blood or organ donors to non-endemic countries [70].

The etiological agent of CD is *Trypanosoma cruzi*, a flagellated protozoan of the order Kinetoplastida, family Trypanosomatidae, genus Trypanosoma, which is characterised by the existence of a flagellum and a single mitochondrion, where the kinetoplast - a specialised organelle containing the DNA - is located. *T. cruzi* is the only human trypanosome transmitted by the excreta of an invertebrate vector and is included in the section Stercoraria. The strains of the parasite circulating in nature,

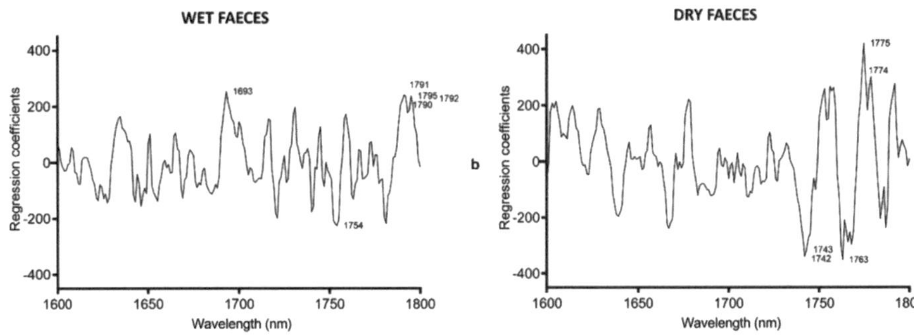

Figure 7.
Regression coefficients using 6 and 5 factors for (a) wet and (b) dry excreta, respectively, in the PLS model based on the NIR spectra in the 1600–1800-nm region for differentiating T. cruzi-infected from uninfected excreta. Adapted from [73].

during their life cycle, can infect different vertebrates (humans, domestic animals and other wild mammals) and invertebrate (triatomines) hosts. Throughout its life cycle, *T. cruzi* assumes morphologically and physiologically distinct evolutionary forms. These evolutionary forms are identified by the relative position of the kinetoplast in relation to the cell nucleus and the appearance of the flagellum, the main ones being: trypomastigote, amastigote and epimastigote [71].

The vectors of CD are insects of the Triatominae subfamily (Hemiptera: Reduviidae) and are popularly known as *barbeiros* (Brazil), *vinchuca* (Argentina, Bolivia and Chile), *chinchorro* (Ecuador), *chirimacha* (Peru), *chipo* (Venezuela) and kissing bug (USA). They are nocturnal insects a characteristic that helps their hemophagic behaviour, because they take advantage of the resting period of their hosts to perform their blood repast while reducing the chances of being predated. Currently, there are 150 described species of triatomines, all of which are considered potential vectors of CD. The main vector of CD in the Americas is *Triatoma infestans*, although other species such as *T. dimidiata*, *T. brasiliensis*, and *Rhodnius prolixus* among others, may be of high relevance in specific geographic areas [72].

NIRS has only been recently applied to detect *T. cruzi* in Triatomine species reared in the lab. It was 100% accurate for predicting *T. cruzi* infection in either the midgut or the rectum of *T. infestans*. Furthermore, the model developed for predicting infection in insect midgut could be applied for diagnosis of *T. cruzi* on the rectum and vice-versa. The authors also demonstrated that NIRS could accurately predict infection with 100% accuracy using either wet or dry samples [73]. **Figure 7** shows regression coefficients used to differentiate infected from uninfected excreta.

3. Mid infrared for predicting age, species and infection

MIR is a recently developed technique in the field of vector-borne diseases. For age grading mosquitoes, it has been demonstrated on the African malaria mosquitoes *An. gambiae* and *An. arabiensis* and *Aedes aegypti* mosquitoes. The capacity of MIR to predict the age of laboratory-reared mosquitoes was demonstrated by Jiménez and colleagues in 2019 [74] using Bruker Vertex 70 spectrometer (Bruker Corporation, Billerica, Massachusetts, USA). Here they showed that the age of young (1d and 15 d old mosquitoes) could be differentiated from the middle age groups. In their most recent study on age grading, they scaled up the sample size of Anopheles mosquitoes

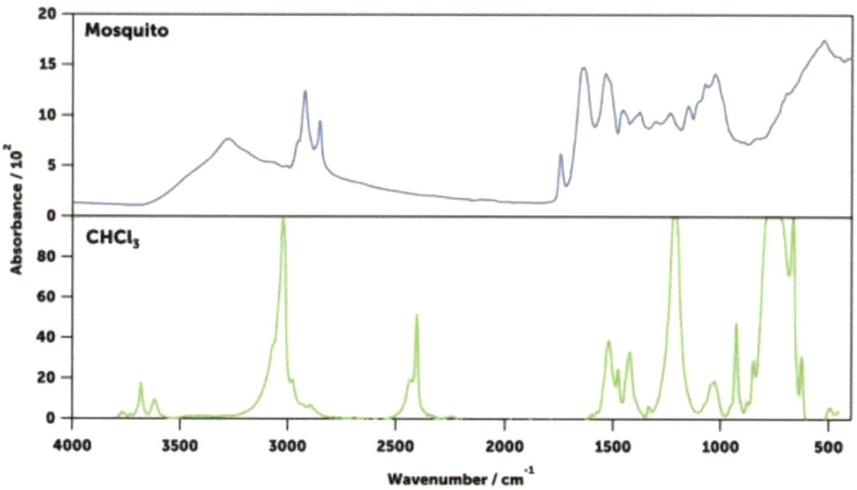

Figure 8.
MIR spectra of a mosquito and chloroform adapted from [74].

to 40,000 and demonstrated the technique's capacity to distinguish mosquitoes with varying egg laying history that is, those with 1–3 gonotrophic cycles [75]. In addition to age grading, MIR has been shown as a predictor of the source of host blood meal of blood-fed Anopheles mosquitoes [76]. This is important as it demonstrates whether a mosquito species prefers one host over the other or if it can feed on multiple hosts species. In addition to age grading and blood meal analysis, MIR can differentiate cryptic species of *An. gambiae* with a predictive accuracy of 90% for both laboratory and field-collected samples [74].

Other studies have reported on the capacity of MIR to detect infections such as *Wolbachia* in mosquitoes [77] as well as a diagnostic tool for Plasmodium parasites in blood samples from human subjects in Thailand [78] and Tanzania [79]. Overall, the findings of these studies demonstrate the potential of MIR technique for multiple applications in the field of mosquito-borne diseases. A raw spectrum of a mosquito collected by mid-infrared is shown in **Figure 8**.

4. Preservation techniques compatible with infrared techniques

Most vector control interventions conduct large scale studies in remote areas where thousands of samples are collected, stored and processed at a later stage in the laboratory. These samples are often stored in preservatives to ensure the DNA/RNA does not degrade prior to processing of the samples. For infrared techniques to be incorporated in such studies, the predictive accuracy of samples should not be affected by these preservatives. Three studies have demonstrated that a number of preservatives are compatible with these techniques including Silica gel, ethanol, Carnoy's solution, RNAlater solution and storage in 4 degrees [78–80]. The predictive accuracy of samples preserved by these techniques was observed from 1 to 52 weeks. Overall, preservation with the silica gel technique has been recommended for both MIR and NIR techniques for periods of 2–52 weeks [72]. RNAlater, storage in 4 degrees was also shown accurately for the longest period that the samples were stored under that is 4 weeks [78]. Based on the nature of studies conducted for mosquito control programs, silica gel is the preferred preservation technique as it is associated with low cost and can used in the field under varying temperatures.

5. Conclusions

The application of both Near and mid-Infrared spectroscopy techniques for age, species and infection detection for malaria and arbovirus vectors is a relatively new area of research. Several studies have reported the potential of infrared techniques as alternative tools for vector surveillance and diagnostic tools. Relative to existing techniques, they are rapid and cost-effective and can easily be scaled up for programmatic use. However, there is very limited data on their applicability in the field. It is recommended that for both techniques to be field ready, robust models should be developed and validated using field collected samples from multiple settings. Secondly, none of the published studies provide evidence whether their ability to predict age, species or infection is based on the detection of specific chemical compounds or the actual pathogen presented in the mosquito/human samples. This needs to be established to improve the predictive accuracy of these techniques. Overall infrared spectroscopy techniques represent a big step forward in the field of vector-borne diseases and have potential to stop outbreaks in a timely manner through rapid identification of hotspots.

Acknowledgements

We would like to thank Ms. Tharanga Kariyawasam (The University of Queensland) for assistance with references and Dr. Gabriela A Garcia (Fiocruz), and Associate Professor Anton Lord (Queensland University of Technology) for their initial contribution to this chapter. We also appreciate the Fundação de Amparo a Pesquisa no Estado do Rio de Janeiro (FAPERJ), Conselho Nacional de Desenvolvimento Científico e Tecnológico (CNPq), Fiocruz (INOVA) and the University of Queensland for funding.

Conflict of interest

The authors declare no conflict of interest.

Author details

Maggy Sikulu-Lord[1*] and Rafael Maciel-de-Freitas[2,3]

1 The School of Biological Sciences, The University of Queensland, St Lucia, Australia

2 Laboratório de Transmissores de Hematozoários, Instituto Oswaldo Cruz, Fiocruz, Rio de Janeiro, Brazil

3 Department of Arbovirology, Bernhard-Nocht Institute for Tropical Medicine, Hamburg, Germany

*Address all correspondence to: maggy.lord@uq.edu.au

References

[1] Runge-Ranzinger S et al. What does dengue disease surveillance contribute to predicting and detecting outbreaks and describing trends? Tropical Medicine & International Health. 2008;**13**(8):1022-1041

[2] Runge-Ranzinger S et al. Dengue disease surveillance: An updated systematic literature review. Tropical Medicine & International Health. 2014;**19**(9):1116-1160

[3] Pley C et al. Digital and technological innovation in vector-borne disease surveillance to predict, detect, and control climate-driven outbreaks. The Lancet Planetary Health. 2021;**5**(10):e739-e745

[4] Braks M et al. Making vector-borne disease surveillance work: New opportunities from the SDG perspectives. Frontiers in veterinary science. 2019:232

[5] Dzul-Manzanilla F et al. Identifying urban hotspots of dengue, chikungunya, and Zika transmission in Mexico to support risk stratification efforts: A spatial analysis. The Lancet Planetary Health. 2021;**5**(5):e277-e285

[6] Daszak P. Global trends in emerging infectious diseases. Nature. 2008;**451**(7181):990-993

[7] Rozendaal JA. Vector Control: Methods for Use by Individuals and Communities. World Health Organization; 1997

[8] Sokhna C, Ndiath M, Rogier C. The changes in mosquito vector behaviour and the emerging resistance to insecticides will challenge the decline of malaria. Clinical Microbiology and Infection. 2013;**19**(10):902-907

[9] Russell TL et al. Increased proportions of outdoor feeding among residual malaria vector populations following increased use of insecticide-treated nets in rural Tanzania. Malaria Journal. 2011;**10**(1):1-10

[10] Walton C et al. Molecular identification of mosquito species. Biological Journal of the Linnean Society. 1999;**68**(1-2):241-256

[11] Beier CJ. Malaria parasite development in mosquitoes. Annual Review of Entomology. 1998;**43**:519-543

[12] Chan M, Johansson MA. The incubation periods of dengue viruses. PLoS One. 2012;**7**(11):e50972

[13] Kuno G. Universal diagnostic RT-PCR protocol for arboviruses. Journal of Virological Methods. 1998;**72**(1):27-41

[14] Tassanakajon A et al. Polymerase chain reaction detection of plasmodium falciparum in mosquitoes. Transactions of the Royal Society of Tropical Medicine and Hygiene. 1993;**87**(3):273-275

[15] Wirtz R et al. Field evaluation of enzyme-linked immunosorbent assays for plasmodium falciparum and plasmodium vivax sporozoites in mosquitoes (Diptera: Culicidae) from Papua New Guinea. Journal of Medical Entomology. 1987;**24**(4):433-437

[16] Sylvestre G et al. Preliminary evaluation on the efficiency of the kit Platelia dengue NS1 Ag-ELISA to detect dengue virus in dried Aedes aegypti: A potential tool to improve dengue surveillance. Parasites & Vectors. 2014;**7**(1):1-7

[17] Abraham PR et al. Detection of recombinant dengue virus 2 NS1

protein in Aedes aegypti mosquitoes using commercial dengue NS1 ELISA kit. Journal of Vector Borne Diseases. 2022;**59**(1):98

[18] Beier JC et al. Malaria sporozoite detection by dissection and ELISA to assess infectivity of afrotropical anopheles (Diptera: Culicidae). Journal of Medical Entomology. 1990;**27**(3):377-384

[19] McMeniman CJ et al. Stable introduction of a life-shortening Wolbachia infection into the mosquito Aedes aegypti. Science. 2009;**323**(5910):141-144

[20] Hoffmann AA et al. Successful establishment of Wolbachia in Aedes populations to suppress dengue transmission. Nature. 2011;**476**(7361):454-457

[21] Aliota MT et al. The wMel strain of Wolbachia reduces transmission of Zika virus by Aedes aegypti. Scientific Reports. 2016;**6**:28792

[22] Walker T et al. The wMel Wolbachia strain blocks dengue and invades caged Aedes aegypti populations. Nature. 2011;**476**(7361):450-453

[23] Lee SF et al. High-throughput PCR assays to monitor Wolbachia infection in the dengue mosquito (Aedes aegypti) and Drosophila simulans. Applied and Environmental Microbiology. 2012;**78**(13):4740-4743

[24] Detinova T. Age-grouping methods in Diptera of medical importance, with special reference to some vectors of malaria. Monograph Series. World Health Organization. 1962;**47**:13-191

[25] Polovodova V. Age changes in ovaries of anopheles and methods of determination of age composition in mosquito population. Med Parazit (Mosk). 1941;**10**:387

[26] Polovodova VP. The determination of the physiological age of female anopheles by number of gonotrophic cycles completed. Medskaya Parazitol Parazitar Bolezni. 1949;**18**:352-355

[27] Perez-Mendoza J et al. Chronological age-grading of house flies by using near-infrared spectroscopy. Journal of Medical Entomology. 2002;**39**(3):499-508

[28] Mayagaya VS et al. Non-destructive determination of age and species of Anopheles gambiae s.l. using near-infrared spectroscopy. The American Journal of Tropical Medicine and Hygiene. 2009;**81**:622-630

[29] Sikulu M et al. Near-infrared spectroscopy as a complementary age grading and species identification tool for African malaria vectors. Parasites & Vectors. 2010;**3**:49

[30] Sikulu MT et al. Using a near-infrared spectrometer to estimate the age of anopheles mosquitoes exposed to pyrethroids. PLoS One. 2014;**9**(3):e90657

[31] Sikulu-Lord MT et al. Near-infrared spectroscopy, a rapid method for predicting the age of male and female wild-type and Wolbachia infected Aedes aegypti. PLoS Neglected Tropical Diseases. 2016;**10**(10):e0005040

[32] Sikulu-Lord MT et al. First report on the application of near-infrared spectroscopy to predict the age of Aedes albopictus Skuse. Scientific Reports. 2018;**8**(1):9590

[33] Milali MP et al. Age grading an. Gambiae and an. Arabiensis using near infrared spectra and artificial neural networks. PLoS One. 2019;**14**(8):e0209451

[34] Milali MP et al. An autoencoder and artificial neural network-based method to estimate parity status of wild mosquitoes from near-infrared spectra. PLoS One. 2020;**15**(6):e0234557

[35] Gerade BB et al. Field validation of Aedes aegypti (Diptera: Culicidae) age estimation by analysis of cuticular hydrocarbons. Journal of Medical Entomology. 2004;**41**(2):231-238

[36] Hugo LE et al. Proteomic biomarkers for ageing the mosquito Aedes aegypti to determine risk of pathogen transmission. PLoS One. 2013;**8**(3):e58656

[37] Sikulu MT et al. Proteomic changes occurring in the malaria mosquitoes Anopheles gambiae and Anopheles stephensi during aging. Journal of Proteomics. 2015;**126**:234-244

[38] Cook PE et al. The use of transcriptional profiles to predict adult mosquito age under field conditions. Proceedings of the National Academy of Sciences. 2006;**103**(48):18060-18065

[39] Reeves WK et al. Age-grading the biting midge Culicoides sonorensis using near-infrared spectroscopy. Medical and Veterinary Entomology. 2010;**24**(1):32-37

[40] Aw WC, Dowell FE, Ballard JWO. Using near-infrared spectroscopy to resolve the species, gender, age, and the presence of Wolbachia infection in laboratory-reared drosophila. G3: Genes| genomes|. Genetics. 2012;**2**(9):1057-1065

[41] Goh B et al. The application of spectroscopy techniques for diagnosis of malaria parasites and arboviruses and surveillance of mosquito vectors: A systematic review and critical appraisal of evidence. PLoS Neglected Tropical Diseases. 2021;**15**(4):e0009218

[42] Fernandes JN et al. Rapid, noninvasive detection of Zika virus in Aedes aegypti mosquitoes by near-infrared spectroscopy. Science Advances. 2018;**4**(5):eaat0496

[43] Gu W et al. Fundamental issues in mosquito surveillance for arboviral transmission. Transactions of the Royal Society of Tropical Medicine and Hygiene. 2008;**102**(8):817-822

[44] Lau SM et al. Surveillance of adult Aedes mosquitoes in Selangor, Malaysia. Tropical Medicine & International Health. 2015;**20**(10):1271-1280

[45] Souza-Neto JA, Powell JR, Bonizzoni M. Aedes aegypti vector competence studies: A review. Infection, Genetics and Evolution. 2019;**67**:191-209

[46] Brady OJ, Hay SI. The global expansion of dengue: How Aedes aegypti mosquitoes enabled the first pandemic arbovirus. Annual Review of Entomology. 2020;**65**:191-208

[47] Weaver SC et al. Zika, chikungunya, and other emerging vector-borne viral diseases. Annual Review of Medicine. 2018;**69**:395

[48] Gaudreault NN et al. African swine fever virus: An emerging DNA arbovirus. Frontiers in Veterinary Science. 2020;**7**:215

[49] Ferreira-de-Lima VH, Lima-Camara TN. Natural vertical transmission of dengue virus in Aedes aegypti and Aedes albopictus: A systematic review. Parasites & Vectors. 2018;**11**(1):1-8

[50] Salazar MI et al. Dengue virus type 2: Replication and tropisms in orally infected Aedes aegypti mosquitoes. BMC Microbiology. 2007;**7**(1):1-13

[51] Musso D, Gubler DJ. *Zika virus*. Clinical Microbiology Reviews. 2016;**29**(3):487-524

[52] Musso D, Desprès P. Serological diagnosis of flavivirus-associated human infections. Diagnostics. 2020;**10**(5):302

[53] Munoz-Jordan JL. Diagnosis of Zika virus infections: Challenges and opportunities. The Journal of Infectious Diseases. 2017;**216**(suppl_10):S951-S956

[54] Silva JV Jr et al. A scoping review of chikungunya virus infection: Epidemiology, clinical characteristics, viral co-circulation complications, and control. Acta Tropica. 2018;**188**:213-224

[55] Reusken CB et al. Laboratory preparedness and response with a focus on arboviruses in Europe. Clinical Microbiology and Infection. 2018;**24**(3):221-228

[56] Ellwanger JH, Kaminski VDL, Chies JAB. How to detect new viral outbreaks or epidemics? We need to survey the circulation of viruses in humans and other animals using fast, sensible, cheap, and broad-spectrum methodologies. Brazilian Journal of Infectious Diseases. 2017;**21**:211-212

[57] Ramírez AL et al. Searching for the proverbial needle in a haystack: Advances in mosquito-borne arbovirus surveillance. Parasites & Vectors. 2018;**11**(1):1-12

[58] Santos L et al. High throughput estimates of Wolbachia, Zika and chikungunya infection in Aedes aegypti by near-infrared spectroscopy to improve arbovirus surveillance. Communications Biology. 2021;**4**(1):1-9

[59] Landmann F. The Wolbachia endosymbionts. Microbiology Spectrum. 2019;7(2) 7.2:25

[60] Taylor MJ et al. W olbachia filarial interactions. Cellular Microbiology. 2013;**15**(4):520-526

[61] Kaur R et al. Living in the endosymbiotic world of Wolbachia: A centennial review. Cell Host & Microbe. 2021;**29**(6):879-893

[62] Hilgenboecker K et al. How many species are infected with Wolbachia?–a statistical analysis of current data. FEMS Microbiology Letters. 2008;**281**(2):215-220

[63] Moreira LA et al. A Wolbachia symbiont in Aedes aegypti limits infection with dengue, chikungunya, and plasmodium. Cell. 2009;**139**(7):1268-1278

[64] Terradas G, McGraw EA. Wolbachia-mediated virus blocking in the mosquito vector Aedes aegypti. Current Opinion in Insect Science. 2017;**22**:37-44

[65] Sikulu-Lord MT et al. Rapid and non-destructive detection and identification of two strains of Wolbachia Aedes aegypti by near-infrared spectroscopy. PLoS Neglected Tropical Diseases. 2016;**10**(6):e0004759

[66] Esperança PM et al. Detection of plasmodium berghei infected Anopheles stephensi using near-infrared spectroscopy. Parasites & Vectors. 2018;**11**(1):1-9

[67] Maia MF et al. Detection of plasmodium falciparum infected Anopheles gambiae using near-infrared spectroscopy. Malaria Journal. 2019;**18**(1):1-11

[68] Da DF et al. Detection of plasmodium falciparum in laboratory-reared and naturally infected wild mosquitoes using near-infrared spectroscopy. Scientific Reports. 2021;**11**(1):1-8

[69] James SL et al. Global, regional, and national incidence, prevalence, and years

lived with disability for 354 diseases and injuries for 195 countries and territories, 1990-2017: A systematic analysis for the global burden of disease study 2017. The Lancet. 2018;**392**(10159):1789-1858

[70] Bern C et al. Chagas disease in the United States: A public health approach. Clinical Microbiology Reviews. 2019;**33**(1):e00023-e00019

[71] Burleigh BA, Andrews NW. The mechanisms of Trypanosoma cruzi invasion of mammalian cells. Annual Review of Microbiology. 1995;**49**:175-201

[72] Monteiro FA et al. Evolution, systematics, and biogeography of the Triatominae, vectors of Chagas disease. Advances in Parasitology. 2018;**99**:265-344

[73] Tátila-Ferreira A et al. Near infrared spectroscopy accurately detects Trypanosoma cruzi non-destructively in midguts, rectum and excreta samples of Triatoma infestans. Scientific Reports. 2021;**11**(1):1-10

[74] Jiménez MG et al. Prediction of mosquito species and population age structure using mid-infrared spectroscopy and supervised machine learning. Wellcome open research. 2019;**4**

[75] Siria DJ et al. Rapid age-grading and species identification of natural mosquitoes for malaria surveillance. Nature Communications. 2022;**13**(1):1-9

[76] Mwanga EP et al. Using mid-infrared spectroscopy and supervised machine-learning to identify vertebrate blood meals in the malaria vector. Anopheles arabiensis. Malaria Journal. 2019;**18**(1):1-9

[77] Khoshmanesh A et al. Screening of Wolbachia endosymbiont infection in Aedes aegypti mosquitoes using attenuated total reflection mid-infrared spectroscopy. Analytical Chemistry. 2017;**89**(10):5285-5293

[78] Heraud P et al. Infrared spectroscopy coupled to cloud-based data management as a tool to diagnose malaria: A pilot study in a malaria-endemic country. Malaria Journal. 2019;**18**(1):1-11

[79] Mwanga EP et al. Detection of malaria parasites in dried human blood spots using mid-infrared spectroscopy and logistic regression analysis. Malaria Journal. 2019;**18**(1):341

Chapter 8

Application of Infrared Spectroscopy in the Characterization of Lignocellulosic Biomasses Utilized in Wastewater Treatment

Marwa El-Azazy, Ahmed S. El-Shafie and Khalid Al-Saad

Abstract

Global economies are confronting major energy challenges. Mitigating the energy depletion crisis and finding alternative and unconventional energy sources have been subjects for many investigations. Plant-sourced biomasses have started to attract global attention as a renewable energy source. Maximizing the performance of the biomass feedstock in different applications requires the availability of reliable and cost-effective techniques for characterization of the biomass. Comprehending the structure of lignocellulosic biomass is a very important way to assess the feasibility of bond formation and functionalization, structural architecture, and hence sculpting of the corresponding structure–property liaison. Over the past decades, non-invasive techniques have brought many pros that make them a valuable tool in depicting the structure of lignocellulosic materials. The current chapter will be focused on the applications of Fourier transform infrared (FTIR) spectroscopy especially in the mid-infrared region in the compositional and structural analysis of lignocellulosic biomasses. The chapter will provide a display of examples from the literature for the application of FTIR spectroscopy in finding the composition of various biomasses obtained from different parts of plants and applied for wastewater treatment. A comparison between biomasses and physically/chemically treated products will be discussed.

Keywords: biomasses, leaves, seeds, peels, wastewater treatment, adsorbents, characterization, Fourier transform infrared (FTIR) spectroscopy, adsorption capacity

1. Introduction

Nowadays, the global economies are confronting two major energy challenges: indemnifying an affordable source of energy and simultaneously realizing the transformation into a clean, efficient, and sustainable energy system [1, 2]. To that end,

IntechOpen

mitigating the energy depletion crisis and finding alternative and unconventional energy sources have been the subject for many investigations. Recently, plant-sourced biomasses have started to attract global attention as a sustainable, untapped, and renewable energy source. The feasibility of conversion of the raw feedstock into liquid fuels and other value-added products via processes such as thermochemical, microbial, or enzymatic treatments is an advantage. On the other hand, the chemical composition of the biomasses is unique and could certainly affect these conversion processes and the nature of the obtained products in different ways [1–6]. Moreover, biomasses are abundantly available and therefore can be regenerated in copious amounts annually. Preponderantly, biomass is a composite mixture of both organic (major) and inorganic (minor) components. Lignocellulose is one of the main wood extractives and is usually a mixture of lignin and the cellulosic carbohydrates (cellulose and hemicellulose). The term "lignocellulose" is generally employed to signify the plant biomass [1]. Lignocellulosic biomasses could therefore have different applications.

One of the most important categories of biomasses is agricultural residues or the waste biomass—a secondary product for the agricultural industry [7]. Representing a major part of the annual overall biomass production, especially in developing countries, agro-waste biomasses epitomize a valuable source of energy and fuel. Yet, several other applications entail the usage of biomass both in the raw format (biomass) or following a thermochemical treatment (biochar or hydrochar). These applications include use of biomass/biochar as an animal feed and fertilizer, for pollutant sequestration for wastewater treatment, and as a substitute for petrochemical-based plastics and, hence, biodegradable packaging [6–16].

Guaranteeing and maximizing the performance of biomass feedstock in different applications require the availability of reliable and cost-effective techniques for structural and compositional characterization of the biomass. For example, comprehending the microstructure of lignocellulosic biomass is a very important way to assess the heat and mass transfer phenomena, the feasibility of bond formation and functionalization, and structural architecture and hence sculpting of the corresponding structure–property liaison, which are essential for designing and developing value-added products. While several "traditional" analytical techniques for biomass characterization are already in use, some of these techniques are time-consuming and require tough chemicals that, in turn, could need further remediation. Moreover, some of these techniques require pre-treatment procedures that also encompass the use of hazardous chemicals. Most importantly, many of these techniques are invasive/destructive. Wet chemical methods such as hydrolysis using sulfates, alkali, gas chromatography, high performance liquid chromatography, nitrobenzene oxidation, and transesterification are among these conventional approaches [16–19]. Consequently, finding an effective alternative approach that is cost-effective, non-laborious, and non-invasive for compositional analysis and characterization of the lignocellulosic biomass is important to improve researchers' insights into the structure–property–performance relationships.

From that point, and over the past decades, non-invasive techniques are attracting the attention of the scientific community. In addition to being non-destructive, these techniques bring many other pros such as minimizing the wastes and hence supporting the method greenness, allowing the repetitive measurements of the same property over time, and hence improving the quality of the output. Moreover, these techniques are cost-effective, user-friendly, and fast. Several imaging and spectroscopic techniques could be classified under this category, including scanning electron microscopy (SEM), X-ray micro-computed tomography (X-ray μCT), and Fourier transform infrared (FTIR) and near-infrared spectroscopies (NIRS) [16, 17].

SEM analysis depends on the detection of the interaction of the gold-coated sample atoms with the high-energy (0−30 kV) accelerated and diffracted backscattered electrons to generate the SEM micrographs that show the morphological and topography data (e.g., crystallinity, variations in chemical structure prior to and following any sort of treatment, fiber distribution, and size variations). In addition, elemental analysis of the surface can also be performed with the energy dispersive X-ray (EDX) aspect. Compared to optical microscopes, SEM has a much higher resolution that could hit 0.01 μm [16, 20–23]. On the other hand, X-ray μCT depends on the creation of virtual 3D images following the interaction of the X-ray photons with the sample. X-ray μCT is a non-invasive approach that is useful in finding the internal architecture of the composite biomaterials, and it could provide both qualitative and quantitative data on the spatial distribution of properties such as porosity, average pore-to-pore distance, internal defects, and structural discontinuity [24, 25].

Vibrational spectroscopy-based techniques (infrared and Raman spectroscopies) are used to measure the molecular vibrations following the absorption of photons. Both techniques give spectral sketches that express the chemical temperament of a sample. Three zones could be recognized in the infrared (IR) region: (1) the far IR (FIR, 400–10 cm^{-1}, 25–300 μm), (2) the mid-IR (MIR, 4000–400 cm^{-1}, 2.5–25 μm), and the near-IR (NIR, 14,000–4000 cm^{-1}, 0.7–2.5 μm) [16, 26–28]. Yet, it is important to point out that each of the three regions has substantial applications in different fields.

FTIR spectrometers are usually used for measurements in near- and mid-IR regions. NIRS is basically based on the interaction of light with the sample in the NIR region. Two major processes could be observed in the NIR region: molecular overtones and combination bands associated with hydrogen bonds. Yet, these spectral bands are of much weaker intensities compared to the fundamental vibrational bands, an issue that causes absence of the distinctive characteristics and anharmonicity [16, 26–29]. The NIR spectrometer is simple, speedy, portable, non-destructive, and of low cost and high throughput. Therefore, there are various applications of NIRS including both qualitative and quantitative investigations [30–36]. Selective absorption of NIR radiation by the principal functionalities in the lignocellulosic materials makes NIRS a powerful approach for studying woods and lignocelluloses and identifying the composition of biomasses, their biochemical properties following the application of certain treatment conditions, and the impact of genetic engineering on lignocellulosic feedstock [37–44]. On the other hand, the mid-IR spectra, unlike the NIR spectra, are completely interpretable, especially the absorbance bands, because of the chemical peak specificity. Applications entailing the mid-IR region are therefore more known [45–50].

The current chapter will be focused on the applications of FTIR especially in the mid-IR region in the characterization of lignocellulosic biomasses. The discussion throughout this chapter would revolve around the impact of the detected chemical composition on the performance and adsorptive capacities of the characterized adsorbents for wastewater treatment applications. Readers will be first introduced to the working principle of FTIR spectroscopy and applications in functional groups determination. Application of FTIR spectroscopy in identification of structural changes of agro-waste materials following physical and chemical treatments will be reviewed. Discussion will further continue to include the classification of agricultural waste materials-based sorbents according to the employed plant part. Comparison among the different waste-derived adsorbents will entail features such as the nature of the water pollutant, adsorption capacity, and proposed sorption mechanism.

2. FTIR in the characterization of lignocellulosic biomasses

In this section, an overview of the working principle of FTIR spectroscopy will be presented. The advantages of application of FTIR spectroscopy in analyzing biomass structure and monitoring changes that occur from physical and chemical treatments will be also discussed.

2.1 Working principle and advantages

In general, FTIR spectroscopy is a technique that is based on the detection of the interaction between a substance that possesses a functional moiety (chemical functional groups) and the IR radiation. Samples can be in any physical form (solid, liquid, or gas). An FTIR spectrometer measures the frequencies at which the sample absorbs and generates unique absorption spectra. The spectrometer measures the intensities of these absorption bands as well. With such a unique spectrum, identification of peaks at different wavenumbers is beneficial for the recognition of a particular chemical constituent by contrasting them with those in a reference library. The intensity of the measured peaks can be utilized quantitatively for finding the concentration of a component. The spectrum is a 2D-plot in which the axes are characterized by the intensity and frequency of the sample absorption.

Attenuated total reflection–Fourier transform infrared (ATR-FTIR) spectroscopy is a technique that has a plethora of applications. The coupling of FTIR to ATR allows the determination of samples in different states using different techniques, for example, mounting of solid or liquid samples on an ATR crystal and mixing the powdered sample with KBr [16, 26, 51]. The conventional ATR-FTIR spectroscopy entails using a single detector and generating a single spectrum per measurement. Such a spectrum is an average signal for the area of the sample under consideration. In ATR-FTIR imaging, however, many spatially resolved spectra are compiled utilizing an array detector [52]. Yet, the combination of the power of the FTIR array detectors with that of the algorithms of the software serves to fortify the sensitivity as well as the accuracy of the technique. Furthermore, the absence of the water background absorbance, being non-destructive, the high signal-to-noise ratio, in addition to the fast-scanning speed compared to the dispersive instruments and applicability to a wide range of samples have widened the applications of FTIR spectroscopy. Add to that the liability for coupling to multivariate statistical techniques, which facilitates obtaining quantitative data and hence building predictive models [53].

2.2 Compositional analysis of biomasses

Biomasses can be classified into four classes according to their origin: (1) non-woody plants, which are rich in starch, cellulosic material, and saccharides; (2) woody plants, which are mainly the source of cellulose, hemicellulose, and lignin; (3) agricultural residues, which are also rich in lignocellulosic materials; and (4) biofluids oleaginous materials, which are sources of fatty acids and esters. These materials represent the raw feed for industrial conversion processes, which could be chemical (catalytic), biochemical (fermentation or enzymatic routes), or thermochemical processes (gasification, pyrolysis, and combustion) [54]. Each of these materials before and after conversion processes possesses a unique chemical structure that facilitates its employment for the designated task and helps its identification using the available analytical procedure.

Further classification of biomasses into subgroups, species, and varieties could be done according to their diversity, sources, and origin. For example, wood and woody biomasses could be classified into soft and hard, for example, stems, leaves, shrubs, barks, and branches. Herbaceous and agricultural biomasses could be annual or perennial, field- or processed-based biomasses from different species such as grasses, flowers, straws, stalks, fibers, shells, husks, pits and stones, and other residues such as cobs, pulps, coir, seeds, grains, and so on. Chemically, biomass is a mixture of organic matter (major) and inorganic matter (minor) comprising a variety of solid and fluid well-associated phases. The organic matter is usually in the solid state and could be non-crystalline (cellulose, lignin, etc.) or crystalline (organic minerals). The inorganic content is also solid and could be crystalline (minerals such as phosphates, carbonates, etc.), semi-crystalline, and amorphous (e.g., glasses, silicates). The fluid matter is mostly inorganic and could be in the fluid, liquid, or gas state [55].

Analyzing the composition and hence comprehending the properties of biomasses either in their raw format or following physical or chemical treatments is therefore needed to guarantee the quality of the materials and the production processes and therefore supporting the execution of both market and environmental regulations. This, in turn, facilitates new products to be created that enhance the value of the biomass and the derived products and accordingly support the concepts of the circular economy. First and foremost, determining the target for which the biomass will be used is the most crucial when finding the appropriate analytical technique. For example, the analysis of oleaginous material content that will be utilized to produce biodiesel needs a different analytical approach from those needed for the analysis of sugarcane to be utilized to produce ethanol or sucrose.

Another example: using FTIR spectroscopy to determine the nature of the functional groups on the surface of the biomass/biochar intended to be used for water treatment applications via adsorption helps maximize the removal efficiency and understanding the sorption mechanism. Focusing on fiber research, ATR-FTIR spectroscopy has been widely used for fast depiction of the cell wall structure with an emphasis on the analysis of pectin and its degree of esterification [56, 57]; cell wall polysaccharides, especially cellulose, hemicellulose, lignin, and starch [58–61]; mapping of the chemical composition of investigated fibers with the purpose of assessing the purity and the structure [62]; and the impact of modification (physical or chemical treatments) on fruits' and vegetables' fiber structures [63–66].

The next few subsections will focus on the applications of FTIR spectroscopy for the identification of functional groups in different plant parts and the impact of thermal treatment on the adsorption performance.

2.2.1 Biomass-based adsorbents

Various investigations have been reported in the literature to find the applications of agricultural waste materials (fruits, vegetables, etc.) for wastewater treatment. More than a few advanced composite materials that are agro-waste-based have been engineered for particular applications in wastewater treatment. Agro-wastes, for instance, leaves, seeds, pruning, peels, shells, pulps, and so on, possess distinct elemental compositions and characteristic structures and morphologies, which in the long run impact the product. These wastes represent a burden on the environment if not properly recycled and reused. Therefore, recycling into value-added products for wastewater treatment has been a subject for many investigations, as will be detailed in the next few subsections.

2.2.2 FTIR analysis of peel-based adsorbents

Among the different agro-waste-derived biosorbents, the waste peels are deemed as promising waste resources for the development of biosorbents, thanks to their high lignocellulosic content and the high carbon and oxygen content as illustrated by their elemental composition. Proximate analysis shows that banana peels, for example, are constituted of mainly hemicellulose (41.38%) and cellulose and lignin to a lower extent [67]. On the other hand, biopolymer compositional analysis shows that peanut shells contain 18.8% of cellulose and 81.2% of lignocellulose [68]. **Table 1** shows examples of the applications of peel-based adsorbents (raw, thermally, and chemically treated) for the removal of various pollutants (heavy metals, dyes, and drugs) [69–76]. **Table 1** shows the most characteristic FTIR peaks of the studied peels; the suggested removal mechanism of the target pollutant, whether physi- or chemisorption; as well as the reported maximum sorption capacity (q_{max}, mg/g).

The structural changes in hemicelluloses, cellulose, and lignin following the pyrolysis, torrefaction, or any thermal treatment can be observed in the FTIR spectra. Comparing the spectrum of raw pistachio nutshell (RPNS) to that of the thermally treated candidate at 500°C (PNS500), it can be deduced that the thermal treatment

Adsorbent	Adsorbate	$\nu(cm^{-1})$	Assignment	Mechanism	q_{max} (mg/g)	
Watermelon rinds (thermal treatment)	Acridine orange (AO)	2916	C–H symmetric and asymmetric stretching vibrations	- At low [AO]: Chemisorption – At high [AO]: Physisorption	69.44	[14]
Pistachio nut shells (thermal treatment)	Basic fuchsin	1600	C=C stretching	Physisorption	118.2	[69]
		1379	methyl group vibration			
		1060	C=O vibration of carboxylic acid and ketone			
		1590	–C=C stretching			
		1027	C – O – C stretching for hemicellulose and cellulose			
		1450–1310	Bending vibrations of O-H and C-H groups			
Pomegranate peels (thermal treatment)	Ni(II)	3348	alcohol, phenol, carboxylic	NS*	NS*	[70]
		1227	Bending of –OH			
		3254	N–H stretching			
		2921 & 2879	stretching aliphatic C–H group bond			
		1724	Carbonyl (C=O) stretching vibration from aldehyde, ketone, carboxylic acid			

Adsorbent	Adsorbate	ν(cm^{-1})	Assignment	Mechanism	q_{max} (mg/g)	
		1608.74	Stretching vibrations bonds of C=O and C=C aromatic, C=N, N–H of amines or amides, or C–C aromatic stretching			
		1320	C–O group from alcohol, phenol, ether, ester			
		1017.33	C − O or C–N groups			
		881	C-S bond			
Potato peels (thermal treatment)	Cd(II)	3000–3300	–OH vibration (derived from cellulose, hemicellulose, and lignin)	Physisorption	239.64	[71]
		2938	C-H stretching			
		1632	–C=O stretching vibration			
		1572	aromatic C=C stretching mode			
		1409 & 1154	aryl OH groups			
		1081 & 1364	–OH and C–O stretching			
ZnO-impregnated activated carbon prepared from jackfruit peel	Ciprofloxacin	969.07–1517.25	O–H hydroxyl and C–O–C ether groups in polysaccharides	NS*	15.75	[72]
		1638.39 & 1771.87	C–O &C=O vibration (quinolone)			
		2476.06	COO stretching			
		3414.34	N–H stretching			
Garlic peel (one-step carbonization method)	Enrofloxacin	1675	–COOH group	Physisorption	150.17	[73]
		1514–1465	aromatic C=C			
		1205–1109	–S=O/C–O vibration			
		791	S–O stretching vibration			
Pomelo fruit peel (thermal treatment)	Pb(II)	3449	Stretching vibration of the bonded hydroxyl group in the cellulose molecule	Electrostatic interactions & ion exchanges & adsorptive Interactions	92.13	[74]
		1587	Stretching vibration of the C=C bond			
		1375	Bending vibration of C–H bond			
		1034	Stretching vibration of the C–O bond			

Adsorbent	Adsorbate	ν(cm^{-1})	Assignment	Mechanism	q_{max} (mg/g)	
Persimmon peel (chemical modification)	Ga(III)	1161 and 1726	C=O and C–O stretching	Electrostatic adsorption and cation exchange	128	[75]
		1029	Si–O vibrations			
		3393	–OH vibration			
Banana blossom peels (chemical treatment)	Cr(VI)	3500–3300	Hydroxyl (O-H) stretching vibration of cellulose	NS*	227.27	[76]
		2920	C–H stretching			
		1616	O–H bending of adsorbed water			
		1311 and 1375	C–H & O–H bending vibrations			
		1244	C–O stretching			

**NS: not stated.*

Table 1.
Peel-based adsorbents (raw, thermally treated, chemically modified).

has led to the removal of hydroxyl, carboxylic, carbonyl, and esters groups in the form of water, CO, CO_2, and small volatile organic compounds. Such a behavior could be ascribed to the dehydration that happens during the thermal treatment process (**Figure 1**) [69]. Similar attributes could be observed upon chemical modification of biomasses or biochars [72]. FTIR spectroscopy identifies not only the structural alterations of the lignocellulosic biomass but also the presence of carbohydrate, protein, and lipid in the non-lignocellulosic biomass [77].

Different mechanisms have been reported for the removal of inorganics and organics from wastewater using agro-waste materials. Reflecting on the adsorption mechanism of peel-based sorbent, we can see the involvement of surface complexation, ion-exchange, surface precipitation, physisorption, chemisorption, or both physi- and chemisorption depending on pollutant concentration. Proposal of a sorption mechanism is greatly dependent on the source biomass, the nature of the surface in terms of porosity, surface area, and the existence of functional groups. The presence of polar organic groups on the sorbent surface could help in the formation of chelates with metal-ion pollutants. The role and relevance of functional moieties on the surface of the agroforestry-based sorbents in the removal of organics were confirmed using statistical physics and density functional theory. Findings showed that the existence of hydrogen and oxygen functional groups on the biomass surface were the main responsible functionalities for dye adsorption [78].

2.2.3 FTIR analysis of leaf-based adsorbents

Similar to the peel-based biosorbents, the leaf-based biosorbents are composed of a variety of inorganic and organic compounds, which, in turn, represent the potential

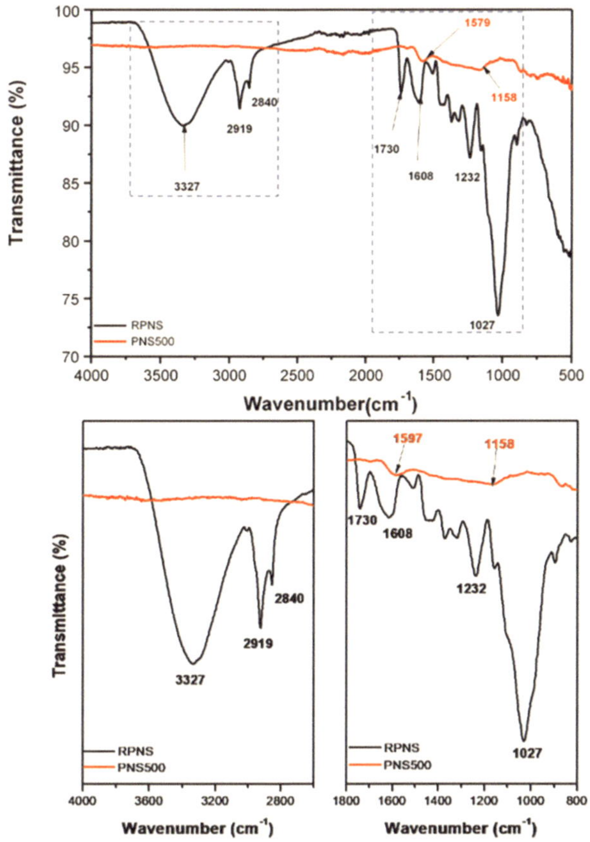

Figure 1.
Fourier-transform infrared (FTIR) spectra of raw pistachio nutshells (RPNS) and the thermally treated sample at 500°C (PNS500) (reproduced from [69], Licensee MDPI, Basel, Switzerland, under the terms and conditions of the Creative Commons Attribution (CC BY) license).

binding sites that could scavenger potential contaminants during the adsorption process. The presence of these potential binding sites is dependent on the experimental conditions and whether the leaves will be used in their raw format (biomass) or after chemical or physical treatments. **Table 2** shows examples from the literature of different leaf-derived biosorbents and their structural composition as elucidated by FTIR spectroscopy [79–85].

The lignocellulosic composition could vary from one plant part to another. In banana leaves, for example and when compared to the peels of the same fruit, reports show that hemicellulose content is much less (23.46%), while the cellulose content is higher (35.58%). The elemental composition of both is almost the same [67]. Pineapple leaves (PAL) have been used for the depollution of rose Bengal-contaminated water [79]. FTIR spectra of the non-thermally treated biomass (RPAL) and the thermally treated species at 250°C (TTPAL250) were depicted and compared. The lignocellulosic composition was reflected on the FTIR spectra with a difference in intensity and location of some peaks. This could be ascribed to the decomposition of lignocellulosic materials upon thermal treatment, a finding that explains the inferior adsorption capability of the thermally treated PAL compared to the RPAL.

Adsorbent	Adsorbate	$\nu(cm^{-1})$	Assignment	Mechanism	q_{max} (mg/g)	
Green tea (thermal treatment)	Methylene blue	3279	–OH vibration or N-H	RGTW: Chemisorption TTGTW500: Physisorption	RGTW: 68.28 TTGTW500: 69.01	[13]
		2918 & 2845	–C–H stretching of the alkane			
		1624	–C=O stretching vibration			
		1533	Secondary amine			
		1455	N–H bending			
		1342	C– H or $-CH_3$ bending			
		1234	$-SO_3$ stretching			
		1146	C – O group			
		1027	C=O			
Pineapple leaves (thermal treatment)	Rose Bengal	3325 & 3318.4	–OH or N–H vibration	Physisorption	58.8	[79]
		2913–2920	C–H stretching			
		1595–1585.8	bending N–H of amines			
		1365 & 1375	Bending–OH			
		1034.3	C–O stretching			
Aloe Vera leaves (thermal treatment)	Titan yellow	3300	–OH vibration	Physisorption	55.25	[15, 80]
		2916–2850	C–H stretching			
		1731	–C=O stretching vibration			
		1586	$-COO^-$ stretching vibration			
		1250	C–C stretching			
		1153	C – O – C stretching of aliphatic ether			
		1019	C–O stretching			
Tea leaf-based biochar (thermal treatment)	Chlortetracycline	3442	Stretching vibrations of O–H	π–π interaction	627	[81]
		1734	C=O group			
		1232	C–O group			
		1060	C–O–C group			
Olive tree leaves (reflux extraction method)	Pb(II)	3500 & 3100	–OH stretching vibration	*NS	35.97	[82]
		2950	Symmetric C–H band			

Adsorbent	Adsorbate	$\nu(cm^{-1})$	Assignment	Mechanism	q_{max} (mg/g)	
		2700	Asymmetric C–H bands			
		1600 & 1650	–OH and –COOH			
		1400–1325	C–O stretching of cellulose			
		1120–950	Organic phosphates (P–O–P), carbonates (C–O), and silicates (Si–O–Si) deformation vibrations			
		780	C–O stretching band			
		568	C–C stretching band			
Platanus orientalis Linn (POL) leaves (modified with $KMnO_4$ & $K_2Cr_2O_7$)	Cd(II)	3700–3200	–OH stretching vibrations	Cation-π interaction, and ion exchange	52.5	[83]
		1750–1540	C=O stretching vibration			
		1420	CO_3^{-2}			
		880	C–H in the benzene rings			
		700–500	Tetrahedral complexes (Mn^{2+}–O^{2-})			
Aegle marmelos leaves (raw material)	Cr(VI)	3275.54	O–H stretch	*NS	8.12	[84]
		2918.58	C–H stretch			
		1595.10	C–C stretch			
		1375.02	N-O asymmetric stretching			
		1243.45	C–N stretch (primary cause of interaction with anions $HCrO_4^-$)			
Mulberry leaves (microwave treatment)	Methyl orange	3680–3000 & 1421	C–H deformation	*NS	MPC dosage (0.1 g) 100: 0.0181	[85]
		1700	Stretching vibration of C=O			
		1980	C=C bonding			
		2882	Lignin composition			
		875	Hemicellulose			

**NS: not stated; RGTW: Raw green tea waste; TTGTW500: Thermally treated green tea waste at 500°C.*

Table 2.
Leaf-based adsorbents (raw, thermally treated, chemically modified).

2.2.4 FTIR analysis of seed-based adsorbents

By the same token, seeds and stones have also been applied to cleanse polluted water. As could be understood, stones and seeds are rich in lignin, and consequently, adsorbents with carbon content of 45–50% could be obtained from seeds and stones of several fruits (**Table 3**) [78, 86–92]. Olive stones have been widely applied for wastewater treatment. Pristine and magnetic stones were applied for the treatment of antibiotic-contaminated wastewater [89]. Peaks attributed to the aromatic skeletal vibration of lignin and C–H deformation of cellulose were detected. Brazilian berry seeds were used to remove methylene blue from wastewater [78]. Results showed that dye could be adsorbed via the interaction with two functional groups of the Brazilian berry seed. FTIR characterization together with modeling and density functional theory showed that the hydrogen and oxygen functionalities could be the main functional groups accountable for the interaction with the dye molecule.

Adsorbent	Adsorbate	$\nu(cm^{-1})$	Assignment	Mechanism	q_{max} (mg/g)	
Date pits (thermal treatment)	Cu(II)	3330	-OH or -NH or both	Mixture of physisorption & chemisorption	4.036	[86]
		2921 & 2843	Aliphatic C–H stretching			
		1739	–C=O stretching vibration			
		1602	–C=C– stretching vibration			
		1039	C − O stretching vibration			
Date pits biochar and Magnetic date pits (thermal treatment followed by chemical modification)	Tigecycline	3181	hydroxyl (–OH) stretching vibration	Chemisorption at low concentrations followed by physisorption at higher drug concentrations	57.14	[87]
		1610 &1632	N–H bending vibration of the quinolines moiety			
		1118 & 1097	C–O stretching of aliphatic ether			
		892 &793	C=C bending of the alkene			
		570	Fe–O bond vibration			
Spent coffee grounds impregnated with TiO_2 (thermal treatment followed by chemical modification)	Balofloxacin	3400	Hydrogen bonded OH	Physisorption	196.73	[88]
		2927 & 2855	aliphatic C–H stretching vibration			
		2164 & 2168	C–C stretching vibration in alkyne			

Adsorbent	Adsorbate	ν(cm^{-1})	Assignment	Mechanism	q_{max} (mg/g)	
		1500–1700	Carboxyl C=O and aromatic C=C stretching vibrations			
		1063	C–OH vibration			
		1630	bending vibration modes of Ti–OH			
		1383	Ti–O stretching mode			
Pristine and magnetic olive stones biochar (thermal treatment followed by chemical modification)	Clofazimine	1580	Aromatic skeletal vibration in lignin	Physisorption	174.03	[89]
		1370	C–H deformation			
		1170	C–O–C vibration			
		890	C–H deformation in cellulose			
		760	Aryl C–H or the aryl C–O groups			
		564	Fe–O bond vibration			
Adenanthera *Paronina* L seed-activated carbon (thermal & chemical modification)	Reactive Red 195 Dye	3602–3617	–OH stretching belongs to alcohols	NS	82.35	[90]
		3413–3435	–NH stretching belongs to aliphatic primary amines			
		2893 & 2924	Aliphatic C–H stretching vibration			
		1380–1384	–CH bending of alkenes			
Sodium alginate–flax seed ash beads (thermal & chemical modification)	Methylene blue	3234, 1594, &1405	Asymmetric and symmetric C=O stretching vibrations	Ion-exchange mechanism	333.3	[91]
		1025	C–O stretching			
		3113	–OH vibrations			
		1593, 1417 & 1015	Asymmetric COO^- symmetric COO^- & C–O–C stretching, respectively			
		1421 and 1030	C=O bending and C–O stretching vibrations			

Adsorbent	Adsorbate	$\nu(cm^{-1})$	Assignment	Mechanism	q_{max} (mg/g)	
Magnetic Fe_3O_4 nanoparticle-loaded papaya (*Carica papaya* L.) seed	Congo red	3439	Stretching vibrations of the hydroxyl group	Hydrogen bonding and electrostatic interactions	216.9	[92]
		2893	C–H stretching vibration in CH and CH_2			
		1638	–C=O stretching vibration			
		1412 &1324	–C–O group			
		1017	Stretching vibrations of the C–O group			
		778	C–H in out-of-plane bending vibrations			
		608	Stretching vibration of Fe–O			
Brazilian berry seeds	Methylene blue	3428	O–H elongation	Hydrogen and oxygen functionalities of biomass surface were the main functional groups responsible for dye adsorption	206.24	[78]
		2925 & 2849	Vibrational bonds of C–H			
		1700	C=O and N–H stretching vibration			
		1061 & 1012	Vibration bonds of C–O			
		602	C–N bond			

Table 3.
Seed-based adsorbents.

2.3 Functional groups and adsorption mechanism

Several variables influence the adsorption mechanism of any adsorbent, including the existence of functional groups, surface area, surface charge, and pH. Several investigations have described the significance of the existent functional groups in removing various contaminants (organics and inorganics) [15, 70, 93]. Chemical interactions of the existing functionalities on the sorbent surface and the different pollutants depend mainly on the carbon surface's chemistry and heterogeneity, the aqueous solution ionic properties, and the pollutant structure. As illustrated in **Figure 2**, the interaction mechanisms between the adsorbate and the adsorbent from an aqueous solution include several mechanisms [94]. Their particular involvement in the adsorption approach differs greatly depending on the adsorbate structure, solution ionic conditions, and the nature of the carbon materials. Two mechanisms are commonly used to describe the adsorption of different pollutants from aqueous solution, including chemisorption (ion exchange, complexation, precipitation) or physisorption (electrostatic interaction) [15, 86–88]. Most of these mechanisms are

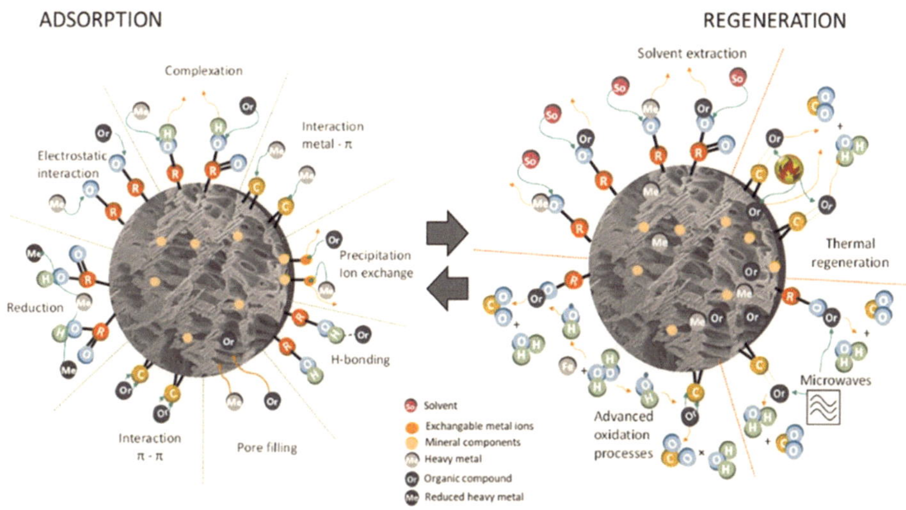

Figure 2.
Adsorption mechanisms onto biochar and their regeneration processes. (Reproduced from [94], Licensee MDPI, under the terms and conditions of the Creative Commons Attribution (CC BY 4.0) license).

strongly related to surface functional groups, such as ion exchange, electrostatic interaction, surface complexation via electrostatic forces, binding site creation, and covalent bonding [95]. One of the key mechanisms for the adsorption of various pollutants, such as heavy metals, is the ion exchange mechanism between pollutants and protons present in oxygen-containing functional groups such as carboxyl and hydroxyl groups [96]. The efficacy of the ion exchange method in the adsorption of various contaminants onto carbonaceous surfaces is mostly determined by the pollutant's ionic size and the kind of functional group of the adsorbent. On the other hand, electrostatic interaction occurs between positively charged pollutants like heavy metals and cationic dyes and negatively charged carbonaceous adsorbents containing anionic functional groups [15]. This process is considered relatively inefficient; the contribution of electrostatic interaction to cationic pollutants' adsorption onto carbonaceous adsorbents is considered a secondary contribution. The reason for that is that the charge of the carbonaceous surfaces is varied, and the efficiency of the electrostatic interaction is dependent mainly on the pH of the adsorbate solution and the point of zero charges of the prepared adsorbent. The interaction between carbon and the pollutant depends on the surface functional groups' ionization degree. Finally, surface complexation mainly describes heavy metals' adsorption and includes the formations of multi-atom structures with particular metal-functional group interactions. For instance, heavy metal can be effectively adsorbed by forming a complex with different functional groups on the adsorbent surface, such as carboxyl, phenolic, and lactone groups [97].

3. Conclusions

Plant-sourced biosorbents are currently attracting global attention. Being easily functionalized with high surface area and superior adsorption capacity for various contaminants, biomasses represent a sustainable solution with a tremendous potential for wastewater treatment. Recycling of these waste materials into value-added products offers ample advantages in terms of protecting the environment from a possible burden in case these wastes were not properly reused and simultaneously offering a green and economic solution for decontaminating water. Analysis of the composition and hence comprehending the properties of biomasses either in their raw format or following physical or chemical treatments is crucial to guarantee the quality of the produced materials and the production processes and therefore supporting the execution of both market and environmental regulations. Throughout this chapter, we have shown the readers the importance of using Fourier transform infrared (FTIR) spectroscopy in depicting the composition of biomass-based adsorbents, an issue that is essential for designing and developing the value-added products. Compared to the traditional analytical techniques, FTIR spectroscopy offers many advantages and is an effective alternative that is cost-effective, non-laborious, and non-invasive for compositional analysis and characterization of the lignocellulosic biomasses.

Conflict of interest

The authors declare no conflict of interest.

Author details

Marwa El-Azazy*, Ahmed S. El-Shafie and Khalid Al-Saad
Department of Chemistry and Earth Sciences, College of Arts and Sciences, Qatar University, Doha, Qatar

*Address all correspondence to: marwasaid@qu.edu.qa

References

[1] El-Azazy M, Bashir S, Liu JL, Shibl MF. Lignin and lignocellulosic materials: A glance on the current opportunities for energy and sustainability. In: Yj G, Song W, Liu JL, Bashir S, editors. Advances in Sustainable Energy. Berlin/Heidelberg, Germany: Springer Cham; 2021. pp. 621-652. DOI: 10.1007/978-3-030-74406-9_22

[2] Chadwick DT, McDonnell KP, Brennan LP, Fagan CC, Everard CD. Evaluation of infrared techniques for the assessment of biomass and biofuel quality parameters and conversion technology processes: A review. Renewable and Sustainable Energy Reviews. 2014;**30**:672-681. DOI: 10.1016/j.rser.2013.11.006

[3] Xiao L, Wei H, Himmel ME, Jameel H, Kelley SS. NIR and Py-mbms coupled with multivariate data analysis as a high-throughput biomass characterization technique: A review. Frontiers in Plant Science. 2014;**5**:388. DOI: 10.3389/fpls.2014.00388

[4] Wiselogel AE, Agblevor FA, Johnson DK, Deutch S, Fennell JA, Sanderson MA. Compositional changes during storage of large round switchgrass bales. Bioresource Technology. 1996;**56**:103-109. DOI: 10.1016/0960-8524(95)00171-9

[5] Sanderson MA, Agblevor F, Collins M, Johnson DK. Compositional analysis of biomass feedstocks by near infrared reflectance spectroscopy. Biomass and Bioenergy. 1996;**11**(5): 365-370. DOI: 10.1016/S0961-9534(96) 00039-6

[6] Osman AI, Fawzy S, Farghali M, et al. Biochar for agronomy, animal farming, anaerobic digestion, composting, water treatment, soil remediation, construction, energy storage, and carbon sequestration: A review. Environmental Chemistry Letters. 2022;**20**:2385-2485. DOI: 10.1007/s10311-022-01424-x

[7] Ullah K, Kumar Sharma V, Dhingra S, Braccio G, Ahmad M, Sofia S. Assessing the lignocellulosic biomass resources potential in developing countries: A critical review. Renewable and Sustainable Energy Reviews. 2015;**51**: 682-698. DOI: 10.1016/j.rser.2015.06.044

[8] Irfan M, Razzaq A, Chupradit S, Javid M, Rauf A, Farooqi TJA. Hydrogen production potential from agricultural biomass in Punjab province of Pakistan. International Journal of Hydrogen Energy. 2022;**47**:2846-2861. DOI: 10.1016/j.ijhydene.2021.10.257

[9] An X, Wu Z, Yu J, Cravotto G, Liu X, Li Q, et al. Copyrolysis of biomass, bentonite, and nutrients as a new strategy for the synthesis of improved biochar-based slow-release fertilizers. ACS Sustainable Chemistry & Engineering. 2020;**8**(8):3181-3190. DOI: 10.1021/acssuschemeng.9b06483

[10] Goiri I, Ruiz R, Atxaerandio R, Lavin JL, de Otálora XD, García-Rodríguez A. Assessing the potential use of a feed additive based on biochar on broilers feeding upon productive performance, pH of digestive organs, cecum fermentation and bacterial community. Animal Feed Science and Technology. 2021;**279**:115039. DOI: 10.1016/j. anife edsci. 2021.115039

[11] El-Gendy A, El-Shafie AS, Issa A, Al-Meer S, Al-Saad K, El-Azazy M. Carbon-based materials (CBMs) for determination and remediation of antimicrobials in different substrates:

Wastewater and infant foods as examples. In: Bartoli M, Frediani M, Rosi L, editors. Carbon-Based Material for Environmental Protection and Remediation. London, UK: IntechOpen; 2020. pp. 103-122. Available from: https://www.intechopen.com/chapters/71482. DOI: 10.5772/intechopen.91750

[12] El-Shafie AS, Ahsan I, Radhwani M, Al-Khangi MA, El-Azazy M. Synthesis and Application of Cobalt Oxide (Co_3O_4)-Impregnated Olive Stones Biochar for the Removal of Rifampicin and Tigecycline: Multivariate Controlled Performance. Nanomaterials. 2022; **12**(3):379. DOI: 10.3390/nano12030379

[13] El-Azazy M, El-Shafie AS, Al-Shaikh YB. Green tea waste as an efficient adsorbent for methylene blue: Structuring of a novel adsorbent using full factorial design. Molecules. 2021; **26**(20):6138. DOI: 10.3390/molecules26206138

[14] El-Shafie AS, Hassan SS, Akther N, El-Azazy M. Watermelon rinds as cost-efficient adsorbent for acridine orange: A response surface methodological approach. Environmental Science and Pollution Research. 2021:1-20. DOI: 10.1007/s11356-021-13652-9

[15] El-Azazy M, El-Shafie AS, El-Gendy A, Issa A, Al-Meer S, Al-Saad K. A comparison between different agro-wastes and carbon nanotubes for removal of sarafloxacin from wastewater: Kinetics and equilibrium studies. Molecules. 2020;**25**:5429. DOI: 10.3390/molecules25225429

[16] Singh SS, Lim LT, Manickavasagan A. Imaging and spectroscopic techniques for microstructural and compositional analysis of lignocellulosic materials: A review. Biomass Conversion and Biorefinery. 2020:1-19. DOI: 10.1007/s13399-020-01075-4

[17] Sluiter JB, Ruiz RO, Scarlata CJ, Sluiter AD, Templeton DW. Compositional analysis of lignocellulosic feedstocks. 1. Review and description of methods. Journal of Agricultural and Food Chemistry. 2010;**58**:9043-9053. DOI: 10.1021/jf1008023

[18] Marulier C, Dumont PJJ, Orgéas L, Caillerie D, du Roscoat SR. Towards 3Danalysis of pulp fibre networks at the fibre and bond levels. Nordic Pulp and Paper Research Journal. 2012;**27**:245-255. DOI: 10.3183/NPPRJ-2012-27-02-p245-255

[19] Viguié J, Latil P, Orgéas L, Dumont PJ, du Roscoat SR, Bloch JF, et al. Finding fibres and their contacts within 3D images of disordered fibrous media. Composites Science and Technology. 2013;**89**:202-210. DOI: 10.1016/j.compscitech.2013.09.023

[20] Temiz C. Scanning Electron Microscopy. In: Mhadhbi M, editor. Electron Microscopy. London, UK: IntechOpen; 2022. DOI: 10.5772/intechopen.103956

[21] Prolongo SG, Gude MR, Del Rosario G, Ureña A. Surface pretreatments for composite joints: Study of surface profile by SEM image analysis. Journal of Adhesion Science and Technology. 2010;**24**(11–12):1855-1867. DOI: 10.1163/016942410X507623

[22] Gupta CK, Sachan AK, Kumar R. Utilization of sugarcane bagasse ash in mortar and concrete: A review. Materials Today: Proceedings. 2022;**65**:798-807

[23] Arkharova NA, Suvorova EI, Severin AV, Khripunov AK, Krasheninnikov SV, Klechkovskaya VV. SEM and TEM for structure and

properties characterization of bacterial cellulose/hydroxyapatite composites. Scanning. 2016;**38**(6):757-765. DOI: 10.1002/sca.21325

[24] Di Giuseppe E, Castellani R, Dobosz S, Malvestio J, Berzin F, Beaugrand J, et al. Reliability evaluation of automated analysis, 2D scanner, and micro-tomography methods for measuring fiber dimensions in polymer-lignocellulosic fiber composites. Composites Part A: Applied Science and Manufacturing. 2016;**90**:320-329. DOI: 10.1016/j.compositesa.2016.07.020

[25] Strandberg A, Thyrel M, Skoglund N, Lestander TA, Broström M, Backman R. Biomass pellet combustion: Cavities and ash formation characterized by synchrotron X-ray micro-tomography. Fuel Processing Technology. 2018;**176**:211-220. DOI: 10.1016/j.fuproc.2018.03.023

[26] El-Azazy M. Introductory chapter: Infrared spectroscopy - A synopsis of the fundamentals and applications. In: El-Azazy M, editor. Infrared Spectroscopy - Principles, Advances, and Applications. London: IntechOpen; 2018. Available from: https://www.intechopen.com/chapters/64432. DOI: 10.5772/intechopen.82210

[27] Tsuchikawa S, Ma T, Inagaki T. Application of near-infrared spectroscopy to agriculture and forestry. Analytical Sciences. 2022;**38**:635-642. DOI: 10.1007/s44211-022-00106-6

[28] Balabin RM, Safieva RZ, Lomakina EI. Comparison of linear and nonlinear calibration models based on near infrared (NIR) spectroscopy data for gasoline properties prediction. Chemometrics and Intelligent Laboratory Systems. 2007;**88**(2):183-188. DOI: 10.1016/j.chemolab.2007.04.006

[29] Pasquini C. Near infrared spectroscopy: Fundamentals, practical aspects and analytical applications. Journal of the Brazilian Chemical Society. 2003;**14**:198-219. DOI: 10.1590/S0103-50532003000200006

[30] Roggo Y, Chalus P, Maurer L, Lema-Martinez C, Edmond A, Jent N. A review of near infrared spectroscopy and chemometrics in pharmaceutical technologies. Journal of Pharmaceutical and Biomedical Analysis. 2007;**44**(3):683-700. DOI: 10.1016/j.jpba.2007.03.023

[31] Scafi SHF, Pasquini C. Identification of counterfeit drugs using near-infrared spectroscopy. The Analyst. 2001;**126**(12):2218-2224. DOI: 10.1039/b106744n

[32] Vredenbregt MJ, Blok-Tip L, Hoogerbrugge R, Barends DM, De Kaste D. Screening suspected counterfeit Viagra® and imitations of Viagra® with near-infrared spectroscopy. Journal of Pharmaceutical and Biomedical Analysis. 2006;**40**(4):840-849. DOI: 10.1016/j.jpba.2005.07.048

[33] Blanco M, Coello J, Iturriaga H, Maspoch S, Pérez-Maseda C. Determination of polymorphic purity by near infrared spectrometry. Analytica Chimica Acta. 2000;**407**(1–2):247-254. DOI: 10.1016/S0003-2670(99)00828-4

[34] Ncama K, Magwaza LS, Mditshwa A, Tesfay SZ. Application of visible to near-infrared spectroscopy for non-destructive assessment of quality parameters of fruit. In: El-Azazy M, editor. Infrared Spectroscopy - Principles, Advances, and Applications. London, UK: IntechOpen; 2018. DOI: 10.5772/intechopen.80069

[35] Freitas MP, Sabadin A, Silva LM, Giannotti FM, do Couto DA, Tonhi E, et al. Prediction of drug dissolution profiles from tablets using NIR diffuse reflectance spectroscopy: A rapid and nondestructive method. Journal of Pharmaceutical and Biomedical Analysis. 2005;**39**(1–2):17-21. DOI: 10.1016/j.jpba.2005.03.023

[36] Broad NW, Jee RD, Moffat AC, Eaves MJ, Mann WC, Dziki W. Non-invasive determination of ethanol, propylene glycol and water in a multi-component pharmaceutical oral liquid by direct measurement through amber plastic bottles using Fourier transform near-infrared spectroscopy. The Analyst. 2000;**125**(11):2054-2058. DOI: 10.1039/B006789J

[37] Fernández JL, Sáez F, Castro E, Manzanares P, Ballesteros M, Negro MJ. Determination of the lignocellulosic components of olive tree pruning biomass by near infrared spectroscopy. Energies. 2019;**12**(13):2497. DOI: 10.3390/en12132497

[38] Sluiter JB, Ruiz RO, Scarlata CJ, Sluiter AD, Templeton DW. Compositional analysis of lignocellulosic feedstocks. Review and description of methods. Journal of Agricultural and Food. Chemistry. 2010;**58**:9043-9053. DOI: 10.1021/jf1008023

[39] Xue J, Yang Z, Han L, Liu Y, Liu Y, Zhou C. On-line measurement of proximates and lignocellulose components of corn stover using NIRS. Applied Energy. 2015;**137**:18-25. DOI: 10.1016/j.apenergy.2014.09.089

[40] Jeong SY, Lee EJ, Ban SE, Lee JW. Near infrared spectroscopy model for analyzing chemical composition of biomass subjected to Fenton oxidation and hydrothermal treatment. Renewable Energy. 2021;**172**:1341-1350. DOI: 10.1016/j.renene.2020.12.020

[41] de Faria BDFH, Barbosa PS, Roque JV, Carneiro ADCO, Rousset P, Candelier K, et al. Evaluation of weight loss and high heating value from biomasses during fungal degradation by NIR spectroscopy. Fuel. 2022;**320**:123841. DOI: 10.1016/j.fuel.2022.123841

[42] Kasemsumran S, Jungtheerapanich S, Ngowsuwan K, Thanapase W, Miyata S. Near-infrared spectroscopic analysis for rapid evaluation of major chemical components in sugarcane bagasse. Thai Journal of Agricultural Science. 2021;**54**(2):104-115

[43] Blanco M, Peinado AC, Mas J. Analytical monitoring of alcoholic fermentation using NIR spectroscopy. Biotechnology and Bioengineering. 2003;**88**:536-542. DOI: 10.1002/bit.20214

[44] Stackpole DJ, Vaillancourt RE, Alves A, Rodrigues J, Potts BM. Genetic variation in the chemical components of eucalyptus globulus wood. G3: Genetics, Genomes. 2011;**1**:151-159. DOI: 10.1534/g3.111.000372

[45] Mitra C. Mid-infrared spectroscopy and challenges in industrial environment. In: El-Azazy M, editor. Infrared Spectroscopy - Principles, Advances, and Applications. London: IntechOpen; 2018. Available from: https://www.intechopen.com/chapters/63369. DOI: 10.5772/intechopen.80657

[46] Bansal R, Singh R, Kaur K. Quantitative analysis of doxorubicin hydrochloride and arterolane maleate by mid IR spectroscopy using transmission and reflectance modes. BMC Chemistry. 2021;**15**(1):1-11. DOI: 10.1186/s13065-021-00752-3

[47] Pacheco-Londoño LC, Castro-Suarez JR, Galán-Freyle NJ, Figueroa-Navedo AM, Ruiz-Caballero JL, Hernández-Rivera RISP. Mid-infrared laser spectroscopy applications I: Detection of traces of high explosives on reflective and matte substrates. In: El-Azazy M, editor. Infrared Spectroscopy - Principles, Advances, and Applications. London: IntechOpen; 2018. Available from: https://www.intechopen.com/chapters/63369. DOI: 10.5772/intechopen.80657

[48] Pacheco-Londoño LC, Galán-Freyle NJ, Padilla-Jiménez AC, Castro-Suarez JR, Figueroa-Navedo AM, Ruiz-Caballero JL, et al. Mid-infrared laser spectroscopy applications in process analytical technology: Cleaning validation, microorganisms, and active pharmaceutical ingredients in formulations. In: El-Azazy M, editor. Infrared Spectroscopy - Principles, Advances, and Applications. London: IntechOpen; 2018. Available from: https://www.intechopen.com/chapters/63369. DOI: 10.5772/intechopen.80657

[49] Bekiaris G, Koutrotsios G, Tarantilis PA, Pappas CS, Zervakis GI. FTIR assessment of compositional changes in lignocellulosic wastes during cultivation of Cyclocybe cylindracea mushrooms and use of chemometric models to predict production performance. Journal of Material Cycles and Waste Management. 2020;**22**(4):1027-1035. DOI: 10.1007/s10163-020-00995-7

[50] Xu F, Yu J, Tesso T, Dowell F, Wang D. Qualitative and quantitative analysis of lignocellulosic biomass using infrared techniques: A mini-review. Applied Energy. 2013;**104**:801-809. DOI: 10.1016/j.apenergy.2012.12.019

[51] Fan M, Dai D, Huang B. Fourier transform infrared spectroscopy for natural fibres. In: Salih SM, editor. Fourier Transform-Materials Analysis. London: IntechOpen; 2012. pp. 45-68. Available from: https://www.intechopen.com/chapters/37067. DOI: 10.5772/35482

[52] Kazarian SG, Chan KA. ATR-FTIR spectroscopic imaging: Recent advances and applications to biological systems. The Analyst. 2013;**138**(7):1940-1951. DOI: 10.1039/C3AN36865C

[53] Canteri MH, Renard CM, Le Bourvellec C, Bureau S. ATR-FTIR spectroscopy to determine cell wall composition: Application on a large diversity of fruits and vegetables. Carbohydrate Polymers. 2019;**212**:186-196. DOI: 10.1016/j.carbpol.2019.02.021

[54] Vaz S Jr. Analytical techniques for the chemical analysis of plant biomass and biomass products. Analytical Methods. 2014;**6**(20):8094-8105. DOI: 10.1039/C4AY00388H

[55] Vassilev SV, Baxter D, Andersen LK, Vassileva CG, Morgan TJ. An overview of the organic and inorganic phase composition of biomass. Fuel. 2012;**94**:1-33. DOI: 10.1016/j.fuel.2011.09.030

[56] Barros AS, Mafra I, Ferreira D, Cardoso S, Reis A, da Silva JL, et al. Determination of the degree of methylesterification of pectic polysaccharides by FT-IR using an outer product PLS1 regression. Carbohydrate Polymers. 2002;**50**(1):85-94. DOI: 10.1016/S0144-8617(02)00017-6

[57] Bichara LC, Alvarez PE, Bimbi MVF, Vaca H, Gervasi C, Brandán SA. Structural and spectroscopic study of a pectin isolated from citrus peel by using FTIR and FT-Raman spectra and DFT calculations. Infrared Physics & Technology. 2016;**76**:315-327. DOI: 10.1016/j.infrared.2016.03.009

[58] Liu X, Renard CM, Bureau S, Le Bourvellec C. Revisiting the contribution of ATR-FTIR spectroscopy to characterize plant cell wall polysaccharides. Carbohydrate Polymers. 2021;**262**:117935. DOI: 10.1016/j.carbpol.2021.117935

[59] George N, Andersson AA, Andersson R, Kamal-Eldin A. Lignin is the main determinant of total dietary fiber differences between date fruit (Phoenix dactylifera L.) varieties. NFS Journal. 2020;**21**:16-21

[60] Chylińska M, Szymańska-Chargot M, Kruk B, Zdunek A. Study on dietary fibre by Fourier transform-infrared spectroscopy and chemometric methods. Food Chemistry. 2016;**196**:114-122. DOI: 10.1016/j.foodchem.2015.09.029

[61] Kizil R, Irudayaraj J, Seetharaman K. Characterization of irradiated starches by using FT-Raman and FTIR spectroscopy. Journal of Agricultural and Food Chemistry. 2002;**50**(14):3912-3918. DOI: 10.1021/jf011652p

[62] Chatjigakis AK, Pappas C, Proxenia N, Kalantzi O, Rodis P, Polissiou M. FT-IR spectroscopic determination of the degree of esterification of cell wall pectins from stored peaches and correlation to textural changes. Carbohydrate Polymers. 1998;**37**(4):395-408. DOI: 10.1016/S0144-8617(98)00057-5

[63] Szymańska-Chargot M, Chylińska M, Gdula K, Kozioł A, Zdunek A. Isolation and characterization of cellulose from different fruit and vegetable pomaces. Polymers. 2017; **9**(10):495. DOI: 10.3390/polym9100495

[64] Mothé C, de Miranda I. Characterization of sugarcane and coconut fibers by thermal analysis and FTIR. Journal of Thermal Analysis and Calorimetry. 2009;**97**(2):661-665. DOI: 10.1007/s10973-009-0346-3

[65] Chen Z, Hu TQ, Jang HF. Grant E Modification of xylan in alkaline treated bleached hardwood Kraft pulps as classified by attenuated total-internal-reflection (ATR) FTIR spectroscopy. Carbohydrate Polymers. 2015;**127**: 418-426. DOI: 10.1016/j. carbpol.2015. 03.084

[66] Nair RR, Mondal MM, Weichgrebe D. Biochar from co-pyrolysis of urban organic wastes—Investigation of carbon sink potential using ATR-FTIR and TGA. Biomass Conversion and Biorefinery. 2020;**12**: 4729-4743. DOI: 10.1007/s13399-020-01000-9

[67] Kabenge I, Omulo G, Banadda N, Seay J, Zziwa A, Kiggundu N. Characterization of banana peels wastes as potential slow pyrolysis feedstock. Journal of Sustainable Development. 2018;**11**: 14-24. DOI: 10.5539/jsd.v11n2p14

[68] Fermanelli CS, Córdoba A, Pierella LB, Saux C. Pyrolysis and copyrolysis of three lignocellulosic biomass residues from the agro-food industry: A comparative study. Waste Management. 2020;**102**:362-370. DOI: 10.1016/j.wasman.2019.10.057

[69] El-Azazy M, El-Shafie AS, Ashraf A, Issa AA. Eco-Structured biosorptive removal of basic fuchsin using pistachio nutshells: A definitive screening design—based approach. Applied Sciences. 2019;**9**:4855. DOI: 10.3390/ app9224855

[70] El-Azazy M, Kalla RN, Issa AA, Al-Sulaiti M, El-Shafie AS, Shomar B, et al. Pomegranate peels as versatile adsorbents for water purification: Application of box–Behnken design as a methodological optimization approach.

Environmental Progress & Sustainable Energy. 2019;**38**:1-12. DOI: 10.1002/ep.13223

[71] El-Azazy M, El-Shafie AS, Issa AA, Al-Sulaiti M, Al-Yafie J, Shomar B, et al. Potato peels as an adsorbent for heavy metals from aqueous solutions: Eco-structuring of a green adsorbent operating Plackett-Burman design. Journal of Chemistry. 2019;**2019**:1-14. DOI: 10.1155/2019/4926240

[72] Magesh N, Annam Renita A, Siva R, Harirajan N, Santhosh A. Adsorption behavior of fluoroquinolone (ciprofloxacin) using zinc oxide impregnated activated carbon prepared from jack fruit peel: Kinetics and isotherm studies. Chemosphere. 2022;**290**:133227. DOI: 10.1016/j.chemosphere.2021.133227

[73] Zhao Y, Liu X, Li W, Huang K, Shao H, Qu C, et al. One-step synthesis of garlic peel derived biochar by concentrated sulfuric acid: Enhanced adsorption capacities for Enrofloxacin and interfacial interaction mechanisms. Chemosphere. 2022;**290**:133263. DOI: 10.1016/j.chemosphere.2021.133263

[74] Dinh VP, Nguyen DK, Luu TT, Nguyen QH, Tuyen LA, Phong DD, et al. Adsorption of Pb(II) from aqueous solution by pomelo fruit peel-derived biochar. Materials Chemistry and Physics. 2022;**285**:126105. DOI: 10.1016/j.matchemphys.2022.126105

[75] Gao L, Wang L, Cao Y, Li S. Persimmon peel-based ion-imprinted adsorbent with enhanced adsorption performance of gallium ions. Minerals Engineering. 2022;**176**:107354. DOI: 10.1016/j.mineng.2021.107354

[76] Selimin MA, Latif AFA, Er YC, Muhamad MS, Basri H, Lee TC. Adsorption efficiency of banana blossom peels (musa acuminata colla) adsorbent for chromium (VI) removal. Materials Today: Proceedings. 2022;**57**:1262-1268. DOI: 10.1016/j.matpr.2021.10.502

[77] Chen Y-C, Chen W-H, Lin B-J, Chang J-S, Ong HC. Impact of torrefaction on the composition, structure and reactivity of a microalga residue. Applied Energy. 2016;**181**:110-119. DOI: 10.1016/j.apenergy.2016.07.130

[78] Sellaoui L, Franco D, Ghalla H, Georgin J, Netto MS, Dotto GL, et al. Insights of the adsorption mechanism of methylene blue on brazilian berries seeds: Experiments, phenomenological modelling and DFT calculations. Chemical Engineering Journal. 2020;**394**:125011. DOI: 10.1016/j.cej.2020.125011

[79] Hassan SS, El-Shafie AS, Zaher N, El-Azazy M. Application of pineapple leaves as adsorbents for removal of rose bengal from wastewater: Process optimization operating face-centered central composite design (FCCCD). Molecules. 2020;**25**:3752. DOI: 10.3390/molecules25163752

[80] El-Azazy M, Dimassi SN, El-Shafie AS, Issa AA. Bio-Waste Aloe *vera* leaves as an efficient adsorbent for titan yellow from wastewater: Structuring of a novel adsorbent using plackett-burman factorial design. Applied Sciences. 2019;**9**:4856. DOI: 10.3390/app9224856

[81] Chen Y-P, Zheng C-H, Huang Y-Y, Chen Y-R. Removal of chlortetracycline from water using spent tea leaves-based biochar as adsorption-enhanced persulfate activator. Chemosphere. 2022;**286**:131770. DOI: 10.1016/j.chemosphere.2021.131770

[82] Mahyoob W, Alakayleh Z, Abu Hajar HA, Al-Mawla L, Altwaiq AM, Al-Remawi M, et al. A novel co-processed olive tree leaves biomass for lead adsorption from contaminated water. Journal of Contaminant Hydrology. 2022;**248**:104025. DOI: 10.1016/j.jconhyd.2022.104025

[83] Yin K, Wang J, Zhai S, Xu X, Li T, Sun S, et al. Adsorption mechanisms for cadmium from aqueous solutions by oxidant-modified biochar derived from Platanus orientalis Linn leaves. Journal of Hazardous Materials. 2022;**428**: 128261. DOI: 10.1016/j.jhazmat. 2022.128261

[84] Mathai RV, Mitra JC, Sar SK, Jindal MK. Adsorption of Chromium (VI) from aqueous phase using Aegle marmelos leaves: Kinetics, isotherm and thermodynamic studies. Chemical Data Collections. 2022;**39**:100871. DOI: 10.1016/j.cdc.2022.100871

[85] Siraorarnroj S, Kaewtrakulchai N, Fuji M, Eiad-ua A. High performance nanoporous carbon from mulberry leaves (Morus alba L.) residues via microwave treatment assisted hydrothermal-carbonization for methyl orange adsorption: Kinetic, equilibrium and thermodynamic studies. Materialia. 2022;**21**:101288. DOI: 10.1016/j.mtla. 2021.101288

[86] Al-Saad K, El-Azazy M, Issa AA, Al-Yafie A, El-Shafie AS, Al-Sulaiti M, et al. Recycling of date pits into a green adsorbent for removal of heavy metals: A fractional factorial design-based approach. Frontiers in Chemistry. 2019;**7**(552). DOI: 10.3389/fchem. 2019.00552

[87] El-Azazy M, El-Shafie AS, Al-Meer S, Al-Saad KA. Eco-structured adsorptive removal of tigecycline from wastewater: Date pits' biochar versus the magnetic biochar. Nanomaterials. 2021; **11**:30. DOI: 10.3390/nano11010030

[88] El-Azazy M, El-Shafie AS, Morsy H. Biochar of spent coffee grounds as per se and impregnated with TiO_2: Promising waste-derived adsorbents for balofloxacin. Molecules. 2021;**26**:2295. DOI: 10.3390/molecules26082295

[89] El-Azazy M, Nabil I, Hassan SS, El-Shafie AS. Adsorption characteristics of pristine and magnetic olive stones biochar with respect to clofazimine. Nanomaterials. 2021;**11**:963. DOI: 10.3390/nano11040963

[90] Ashokan P, Asaithambi M, Sivakumar V, Sivakumar P. Batch and column mode adsorption studies of reactive red 195 dye using Adenanthera paronina L seed activated carbon. Groundwater for Sustainable Development. 2021;**15**:100671. DOI: 10.1016/j.gsd.2021.100671

[91] Işık B, Uğraşkan V. Adsorption of methylene blue on sodium alginate–flax seed ash beads: Isotherm, kinetic and thermodynamic studies. International Journal of Biological Macromolecules. 2021;**167**:1156-1167. DOI: 10.1016/j. ijbiomac.2020.11.070

[92] Munagapati VS, Wen H-Y, Gollakota ARK, Wen J-C, Shu C-M, Lin K-YA, et al. Magnetic Fe_3O_4 nanoparticles loaded papaya (Carica papaya L.) seed powder as an effective and recyclable adsorbent material for the separation of anionic azo dye (Congo Red) from liquid phase: Evaluation of adsorption properties. Journal of Molecular Liquids. 2022;**345**:118255. DOI: 10.1016/j.molliq.2021.118255

[93] Machida M, Fotoohi B, Amamo Y, Ohba T, Kanoh H, Mercier L. Cadmium (II) adsorption using functional

mesoporous silica and activated carbon. Journal of Hazardous Materials. 2012; **221–222**:220-227. DOI: 10.1016/j.jhazmat.2012.04.039

[94] Fdez-Sanromán A, Pazos M, Rosales E, Sanromán MA. Unravelling the environmental application of biochar as low-cost biosorbent: A review. Applied Sciences. 2020;**10**(21):7810. DOI: 10.3390/app10217810

[95] Miretzky P, Cirelli AF. Hg (II) removal from water by chitosan and chitosan derivatives: A review. Journal of Hazardous Materials. 2009;**167**:10-23. DOI: 10.1016/j.jhazmat.2009.01.060

[96] Dong L, Hou L, Wang Z, Gu P, Chen G, Jiang R. A new function of spent activated carbon in BAC process: Removing heavy metals by ion exchange mechanism. Journal of Hazardous Materials. 2018;**359**:76-84. DOI: 10.1016/j.jhazmat.2018.07.030

[97] Mohan D, Pittman CU, Bricka M, Smith F, Yancey B, Mohammad J, et al. Sorption of arsenic, cadmium, and lead by chars produced from fast pyrolysis of wood and bark during bio-oil production. Journal of Colloid and Interface Science. 2007;**310**:57-73. DOI: 10.1016/j.jcis.2007.01.020

Chapter 9

Infrared Spectroscopy for Detecting Adulterants in Food and Traditional Indonesian Herbal Medicine

Aulia M.T. Nasution and Hery Suyanto

Abstract

Adulteration in food has a detrimental effect on the product's quality, which may result in nutritional deficiency. It can also be harmful, particularly for medicinal herbal products, as certain adulterants are very toxic to the body. It is thus critical to screen and identify the presence of any probable adulterants in food and herbal products in order to ensure the customers' safety. Infrared Spectroscopy is a very viable technique for such purposes, as specific molecules absorb light at specific wavelengths, which correspond to the vibration frequency of the molecule's bonds. Adulterants typically have their own unique molecular fingerprints, which exhibit their own vibrational spectra. On the basis of these principles, IR Spectroscopy is sensitive enough to detect the presence of potentially detrimental as well as harmful compounds added to food and medicinal products. This chapter describes how Infrared Spectroscopy can be beneficially used in detecting the presence of adulterants that are unintentionally or irresponsibly added to food or traditional herbal products. The last one is more of a result of a lack of knowledge and awareness of food adulterants and their deleterious impacts.

Keywords: infrared spectroscopy, adulterants, vibrational spectra, traditional herbal medicine products

1. Introduction

Adulteration is the irresponsible act to downgrading the quality of food products delivered to the market, either by the replacement of some ingredients with other substitute components or by addition of certain component into its ingredients. The substitute components added into food products are then called as "adulterants" [1]. Responsible authorities must guarantee the safety of food products supplied to consumers in order to prevent any possible health hazards caused by foodborne illness or injury. Adulterations potentially occur at any stage in the food product's supply chain, starting from the farm, in the production facility, through distribution system, and eventually all ways to the market where the products are purchased by customers [2–4].

There are numerous possible practices for such reckless adulteration, which can be broadly classified into intentional and unintentional [1, 5]. Adulteration practices can be driven by many motivations, either to get economic gains or to cause harm to society, i.e., public health disasters, economic reasons, or terror-related purposes. According to the United Nations' Committee on World Food Security [6], specific goals related to the previously described issues are the food security, which means that all people, at all times, have physical, social, and economic access to sufficient, safe, and nutritious food that meets their food preferences and dietary needs for an active and healthy life. In a more visualized way, the classification of any efforts and motivations to undermine food security can be clearly depicted as in **Figure 1**.

Adulteration is also found in many medicinal or pharmaceutical products (drugs). The types of adulteration made in the drugs are varied, from replacing the main drug's active ingredients with similar but cheaper substitutes. These substitutes may have similar or weaker strengths or even induce complementary effects that might affect other health problems in their consumers. In some drugs that are sold illegally, a.k.a. street drugs, psychoactive substances that provide mood-altering, stimulant, or sedative effects are added. In these kinds of drugs, psychoactive substances (such as caffeine, paracetamol, amphetamines, cocaine, and morphine) are added to modify their physicochemical or psychological activity [8–12].

Indonesian traditional herbal medicine (a.k.a. as "*Jamu*") is predominantly made from extracted natural herbs, such as roots, bark, flowers, seeds, leaves, and fruits. Traditionally, Jamu is produced in liquid form, and it has been used by the Indonesian people for centuries to maintain health and alleviate illness. Even though Western (standard) medicine is becoming more prominently used in Indonesia, Jamu remains popular both in rural and urban areas, since people believe that it is regarded safer than any chemical synthetized drug. The production of Jamu is done traditionally with knowledge that is inherited from generation to generation. Unfortunately, many of these traditional producers have lack knowledge about how to process their Jamu in a hygienic manner. Often, the marketed products are found to be severely contaminated with bacteria, yeasts, and molds. Meanwhile, nowadays, Jamu is also processed in various forms, such as capsules, tablets, and powders, and is also labeled with many producers' brands. Unfortunately, some irresponsible manufacturers mixed them

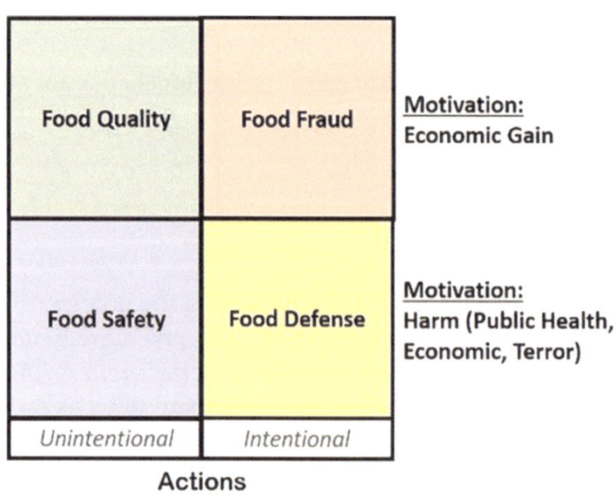

Figure 1.
Food risk matrix (adapted from [7]).

with other chemical drugs as adulterants, which could be toxic and potentially cause health risks [13].

In this chapter, it will be described how Infrared Spectroscopy could beneficially contribute to screening and detecting the presence of adulterants, either in food or Indonesian traditional herbal medicine (Jamu), and how the recognition of adulterants' molecules can be sensed using this measurement technique and the possible recent developments in customizable handheld systems for making the screening task easier.

2. Food fraud and modes of adulteration

The terminology *Food fraud* is described in the first quadrant is defined as any suspected intentional activity by businesses or individuals with intention of deceiving consumers and gaining an economic advantage. These categorized activities are also known as Economically Motivated Adulteration (EMA), which happens when an individual intentionally omits, removes, or replaces a vital element or component of a product. Additionally, EMA occurs when a substance was added to a product with the intention of improving the appearance or value. More specifically, as illustrated in **Figure 2**, there are several forms of food fraud.

Among the seven forms of food fraud, four of them can be considered food adulteration, i.e., dilution, substitution, concealment (masking/hiding), and unapproved enhancements. Meanwhile, the other three, namely counterfeiting, gray market trading, and mislabeling, are all forms of fraud that are not classified as adulteration.

Dilution is often accomplished by increasing the volumetric amount of liquid food or medicinal items by adding additional water or another solvent, thereby reducing the quantity of main ingredients. Sometimes, to conceal the dilution, additional substances such as sugar and artificial flavors are added. The principal reason is economic, since the price of the main ingredients in the diluted product will be lower for the same volume sold, whereas the consumers cannot realize the fraud attempts. One widely practiced example of this kind of fraud is the dilution of pure coconut water with ordinary water and the concealment of the dilution effect with additional sugar [14]. Another common

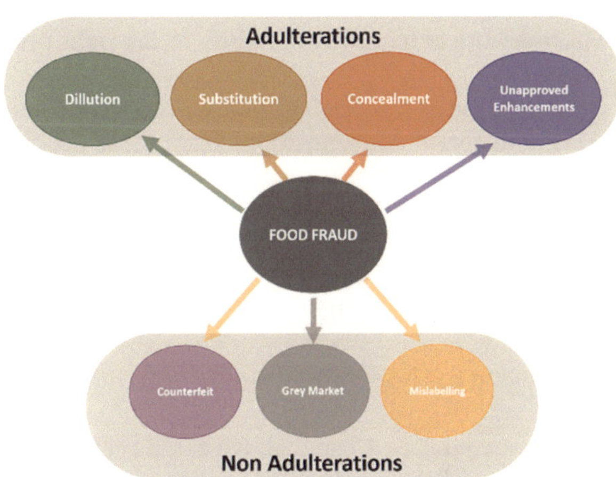

Figure 2.
Numerous forms of food fraud.

example is the dilution of fruit juices, which are considered to be the most targeted food commodities for adulteration and fraud. It is common practice in the fruit juice industry to dilute fruit juice commodities with more water in order to reduce the required fresh fruit and conceal it with the addition of sugars, pulp wash, and other additives such as fruit flavor, as well as the undeclared addition of a lower-quality juice to a product (a.k.a. juice-to-juice adulteration) [15].

Substitution is another common adulteration practice, which is done by replacing the whole or part of the main ingredients with cheaper and harmless substitutes. Milk and dairy products adulteration are good examples of this kind of adulteration practiced in many places on earth. Common substitutes used in milk and dairy products adulteration include more watering, additional vegetable protein, milk from different animal species, whey protein, synthetic urine to boost the nitrogen content, hydrogen peroxide, and even urea and melamine [16]. The last was a well-known Chinese milk scandal that happened in the year 2008 [17].

Unapproved enhancement is the last category of food fraud, which is considered as adulteration. It involves the addition of unknown and unreported substances to food items in order to improve its quality characteristics. This kind of fraud is normally found in many supplemental medicines [18], among others are as dietary and weight-loss supplements [19, 20], cognitive enhancement supplements [21], and supplements for sexual enhancements [12].

3. Fraud and adulteration in medicinal products

Besides being practiced in food commodities, drug fraud and adulteration are also happening in medicinal and pharmaceutical products. Typical adulterants are introduced to raise the product's weight, mimic or enhance its pharmacological action, or facilitate drug transport. Drug adulteration is considered as type of fraud in which legal or illicit drugs are cut or modified to decrease their quality below the level at which they are sold.

There are many facets of pharmaceutical products' counterfeiting that are normally practiced, i.e.,

1. Perfect imitation of the original products, i.e., with the same packaging and the same Active Pharmaceutical Ingredients (APIs) at the right concentration.

2. The packaging is the same as the original pharmaceutical products, but it contains the APIs at different concentrations from those declared.

3. The fake drug looks exactly like the real thing, but it does not have any APIs in it at all.

4. The product contains ingredients different from those declared.

5. The product's packaging has been falsified.

Hence, the points numbered 2, 3, 4 listed above can be considered as adulteration, since it includes modification of the product's ingredients.

Drugs adulteration is harmful and may bring lethal consequence since these adulterated drugs may induce risk of overdose and toxicity that is dangerous to human tissues and organs. The adulterants can be either inert and/or pharmaco-logically active substances that have similar properties to the active ingredients in the original drug itself. These include:

1. diluent substances, i.e., inactive, inert, or structurally different compounds that are added to the drug in order to increase the size of the dosage form and share certain properties such as color, consistency, and taste. Commonly used diluents are anhydrous lactose, lactose monohydrate, talc, and sugar alcohols such as sorbitol, xylitol, and mannitol.

2. adulterants normally used in drugs, among others are:

    a. Caffeine: even often being used as common bulking agents (diluents), caffeine provides stimulant properties similar to cocaine and amphetamine.

    b. Paracetamol: often used as diluents, but also provides analgesic properties similar to heroin.

    c. Lidocaine: has similar anesthetic properties to cocaine.

    d. Procaine: in addition to providing anesthetic properties, also has been found to vaporize heroin at a lower temperature and therefore facilitate smoking

Contamination is another type introduction of unwanted elements into or onto a system during manufacturing, packing, or transport, and the unwanted elements are thus referred to as contaminants. These substances will have an adverse effect on the product or process. Contaminants that are normally found in drugs can be categorized as:

1. Physical contaminants: among others are metal-, glass-, mineral-, and insect fragments, machinery wear debris, greases and oils, dust, and corrosion, fibers, and hairs

2. Chemical contaminants: consist of heavy metals and elemental contaminants; cleaning agents (bleach and detergents); hydrocarbons (oils, gasoline, and diesel); preservatives and colors; and cross-contamination resulting from packaging, manufacturing, or storage.

3. Biological contaminants include viruses, bacteria, and fungi that may be introduced from circulated water and air inside the production system or may be transmitted via the workers that do not strictly follow the Standard Operating Procedures (SOPs).

4. Indonesian herbal medicine and possible adulterants

Indonesia is the world's largest archipelagic country and consists of 18,110 islands with a diverse range of habitats and an extremely complicated geological history similar to neighboring Southeast Asian countries [22]. The climatic, geological, and biodiversity conditions in these regions stimulate the evolution of megadiverse plants with a lot of endemic and ecologically highly adapted species [23]. The biodiversity in Indonesia is considered to be the second largest in the world and has a lot of indigenous medicinal plants. It is thus understandable that many Indonesian communities, particularly in rural areas, still rely on the use of traditional herbal medicines to maintain and improve their health conditions, as well as to prevent and treat diseases [24].

In Indonesia, people generally named these traditional herbal medicines with the word *Jamu*. This word etymologically came from ancient Javanese language, as a contraction of two words, i.e., *jampi* and *usodo*, which means *prayer* and *health*, respectively [25]. The historical records on Jamu as ancient traditional medication were engraved as wall relief sculptures at several Hindu–Buddhist temples such as Borobudur, Prambanan, and Penataran temples, as can be seen in **Figure 3** [25, 26].

Jamu as herbal medications are widely practiced in many communities living in the islands of the Indonesian archipelago, spreading from the main islands of Sumatra

Figure 3.
Historical record on ancient used of Jamu engraved as wall relief sculptures at the Borobudur temple.

[27–29], Java and Madura [30–33], Kalimantan [34–36], Sulawesi [37–39], Papua [40, 41], as well as their smaller neighboring islands. Traditionally Jamu is prepared from raw herbal ingredients in two methods, i.e.:

1. The fresh Jamu (a.k.a. *Jamu segar*): by grating and pounding the raw herbal ingredients using a grater tool or a mortar pounder, respectively, then mixing with fresh drinking water, and squeezing using muslin cloth to extract the herbal juice. The extracted juice is then ready to be consumed.

2. The boiling Jamu (a.k.a. *Jamu godhog*): by boiling mixed herbal ingredients with water in a stewing pot, usually up to the boiling water reduced by half. Upon cooling and filtering, the decoction of herbal extract is ready to be consumed.

These preparation methods are normally adopted as personal consumption in many households or by traditional Jamu's handlers, which can be found in many cities and urban areas in Indonesia. People called these Jamu's handlers according to how they carry their sold products, i.e., as bakul Jamu gendong, bakul Jamu dorong, bakul Jamu sepeda, respectively, as can be depicted in **Figure 4** (i.e., the left, middle, and right).

Meanwhile, in modern processing methods, the Jamu product preparation is done following complex, standardized, and strict methods under controlled standard operating procedures (SOPs). These SOPs are implemented in every processing stage, starting from collecting raw herbal ingredients from farmers, where the originality, purity, water content, and content of active ingredients of each raw herbal component are always checked, since they will contribute to the quality of the final products. Next step is the preprocessing stages including cleaning and washing, chopping into smaller chips of herbal ingredients, drying, and storage for further processing. These preprocessed ingredients are the being weighted and mixed in composition according the prepared prescription, milled, and sieved to get a smooth powdered preparation. This final preparation is then packaged as solid products in capsules, pills, or powdered forms, or being bottled in liquid form upon dilution in prescribed solvents. List of main Jamu factories in Indonesia can be seen in [24].

Either in the traditional or modern processing modes mentioned above, there are open opportunities for adulteration of the prepared products.

Most traditional Jamu handlers are typically honest and humble people who strictly hold the values of responsible business. So, they do not have any intention of

Figure 4.
Typical Jamu handlers in many cities and rural areas in Indonesia.

Figure 5.
Typical street drug handlers that usually sell adulterated and illegal modern processed Jamu.

adulterating their processed and sold products with ingredients other than those taught by their ancestors. Most problems in the products sold by these traditional handlers are solely due to contamination, either chemically or biologically. This contamination is unfortunately caused by a lack of knowledge on how to maintain the hygiene of their raw herbal ingredients, processing tools, and product storage bottles. They do not even understand how to maintain and control the quality of the Jamu they produce. Adulteration in traditional Jamu is only found to be done by a small number of naughty producers and handlers that are hoping to make more profit from their business. They usually use artificial sweeteners such as cyclamate and saccharin [42], which can have negative health effects if consumed for an extended period of time.

Adulteration is also not possible with the Jamu that are produced by main Jamu factories, since their production activities are under the strict supervision of the National Agency of Drug and Food Control (BPOM). Quality screening is regularly done to assure that all licensed factories fulfill the standard of how to make good traditional medicine (abbreviated as "CPOTB" in Indonesian). Factories that are found not to conform to the standard can be further investigated, with the possibility of suspending their operation or even having their production license revoked. Adulteration in modern processed Jamu is often done by unregistered and illegal small home industries, whose adulterated products are usually distributed by street drug handlers, which can be found on many road sides in some areas in big cities, as depicted in **Figure 5**.

5. Infrared spectroscopy: principle of molecule's spectra quantification

Infrared (IR) photon spectra are optical electromagnetic radiation components with their wavelengths ranging from 700 nm to 30 μm. This spectral range is located between visible light and microwave radiation, and the photon's energy of these electromagnetic radiation can be expressed mathematically using a well-known Planck-Einstein equation of $E = h\nu = h(c/\lambda)$. So, the photon energy range of infrared electromagnetic radiation (800 nm–30 μm) spread between 1.78×10^{-1} and

$4.14 \, x \, 10^{-3}$ eV. One electron volt (eV) is defined as the unit of energy required to accelerate an electron through a potential difference of 1 V.

In Infrared Spectroscopy, IR photons are used to probe the spectral absorption behavior of the investigated molecules. When photons of infrared radiation interact with molecules, one possible mode of light-matter interaction is absorption. Different molecules possess specific spectral absorption behaviors that can be used as a determining signature for recognition of the molecule, a.k.a. molecule's spectral fingerprint.

The molecule's spectral absorption profile represents the number of absorbed photons by molecules at different wavelengths of the incoming photons. This spectral absorption behavior is usually represented as percent transmittance (% T) of the infrared radiation that is received by the detector as a function of a photon's wavelength (in nm). But most spectroscopists prefer to represent the absorbance behavior as a function of wavenumber (in cm^{-1}). The units for wavelength and wavenumber are inversely proportional to each other, so spectral range of 700 nm–30 μm correlated to wavenumber of 12,500–333.3 cm^{-1}.

5.1 Molecular vibrations

Atoms in a molecule are connected by chemical bonds having bonding properties determined by their electronegativity (EN), which determines an atom's ability to attract a bonding pair of electrons. An atom's electronegativity is determined by atomic number as well as the configuration of most outer electrons. The difference in electronegativity between two atoms can be used to predict the bonding types between two atoms in a molecule. Based on these differences, chemical bonds between two atoms can be classified into four different classes, i.e.,

- nonpolar covalent bond $(0 < EN < 0.39)$
- moderately polar covalent bond $(0.4 < EN < 0.89)$
- very polar covalent bond $(0.9 < EN < 1.69)$, and
- ionic bond $(EN > 1.7)$

The higher the difference in electronegativity, the stronger the bond between two atoms will be, which means that the higher the required photon energy to break their bond.

The behavior of molecular vibration can be analogically modeled using a harmonic spring-mass system as usually used in explaining well-known Hook's law. The law states that the force F required to extend or compress a spring by some distance Δx is linearly proportional to the distance, with the proportionality constant k, which representing the spring's stiffness. The diatomic model can be seen in **Figure 6**.

When a photon of energy ΔE comes and is absorbed by a molecule, this photon energy will stimulate atoms in a molecule to vibrate, i.e., with the vibration frequency ν according to the following relation, i.e., $\Delta E = h\nu$ where h is the Planck's constant. This photon absorption will stimulate fundamental vibration if the molecule is initially in its ground state. Meanwhile, absorption of multiple photons will stimulate the first or even higher overtones.

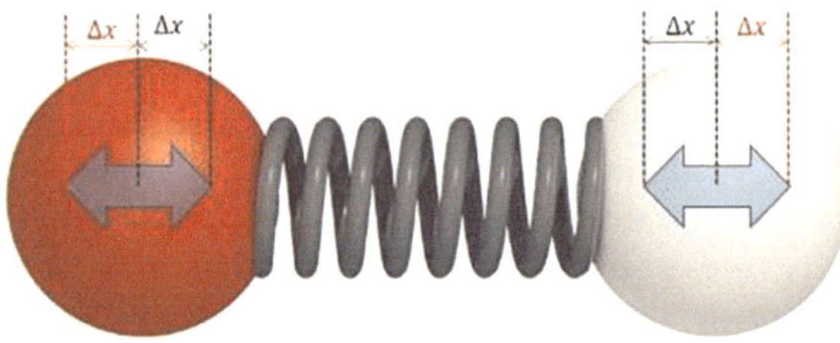

Figure 6.
Analogical harmonic spring-mass system for diatomic molecular vibration, displaced by $\boldsymbol{\Delta x}$*, either compressed (black color) or extended (red color).*

The vibration frequency in wavenumber ($\tilde{\nu}$) can be expressed as:

$$\tilde{\nu}\left[cm^{-1}\right] = \frac{1}{2\pi c}\sqrt{k\frac{(m_1 + m_2)}{m_1 m_2}}(x\ 0.01) \tag{1}$$

Factor 0.01 in the parenthesis is used to convert the unit m^{-1} into cm^{-1}. The vibrational frequency thus depends strongly on the bond's strength, molecular geometry, and the masses of connected atoms. Meanwhile, the bond's strength itself is affected by chemical environment factors and other influencing factors, i.e., hydrogen bonding, coupled vibrations, electronic effects, and Fermi resonance.

5.2 Vibrational modes

As explained previously, the possible vibrational motions in diatomic molecules are compression and extension. Meanwhile, in triatomic or higher polyatomic molecules, the possible motions are more complex. These possible vibrating motions of connected atoms are called vibrational modes, and IR Spectroscopy deals with spectrally quantifying the behavior of these molecular vibrations.

Basically, there are six fundamental (normal) vibrational modes of polyatomic molecules, which can be classified into two classes, i.e., stretching (either symmetrical or asymmetrical) and bending (wagging, twisting, scissoring, and rocking). These bending vibrations can be further divided into two types, i.e., the in-plane (happened on the same plane, i.e., the scissoring and rocking) and out-of-plane (happened not in the same plane, i.e., the wagging and twisting) vibrational modes. Besides these six normal vibrational modes, there are also local modes, which happen as the excitation energies get higher and stimulate the overtones [43, 44].

As an example, let us see the vibrational modes of water molecules, as water is the most common adulterant/diluter that is normally found in food and medicinal products. Water molecule (H_2O) contains three atoms, so it has $3N - 6 = 3(3) - 6 = 3$ vibrations, i.e.:

1. Symmetric stretch ($\tilde{\nu}_1$) at 3280 cm^{-1} (3048.8 nm)

2. Symmetric bending ($\tilde{\nu}_2$) at 1654 cm^{-1} (6045.9 nm)

3. Asymmetric stretch ($\tilde{\nu}_3$) at 3490 cm^{-1} (2865.3 nm)

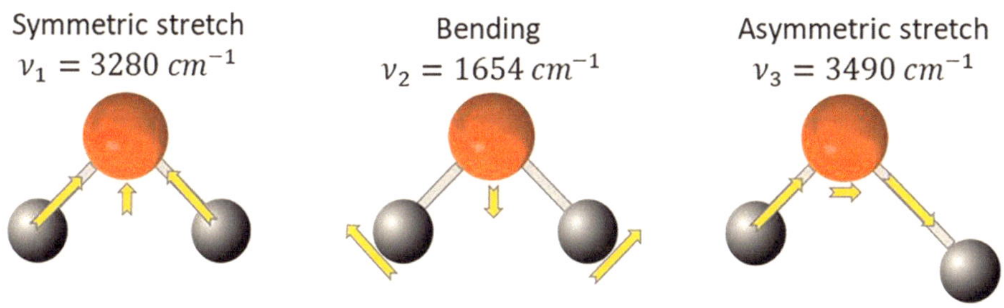

Figure 7.
Normal vibrational modes of water molecules.

These normal modes and related spectrum in the Vis and IR regions can be depicted in **Figure 7**.

5.3 Infrared spectroscopy measurement techniques

There are basically two types of spectrometers used in IR Spectroscopy, i.e., the Dispersive IR (DIR) and the Fourier Transform IR (FTIR) spectrometers. The first typically uses radiation from a broadband source that passes through the sample, which is then dispersed by a monochromator into component frequencies that are further directed onto a detector for recording the spectrum. This recorded spectrum is then compared with the reference beam, as most of the Dispersive IR spectrometers have a double-beam configuration.

Meanwhile, the second type adopts the interferometric measuring principle, where the splitted reference and measuring beams are combined to make an interference pattern (interferogram). This recorded interferogram is then processed by using a Fourier Transform algorithm to extract the spectrum. The FTIR is preferable to the DIR due to superior speed and sensitivity, since all frequencies are examined simultaneously in a shorter time. Detailed comparison of these two measurement techniques can be found in [45].

5.4 Processing and interpretation of IR spectra

In order to obtain accurately identified spectra of examined samples, it is critical to ensure that a number of steps, beginning with sample preparation and measurements, preprocessing, reading, and interpreting the measured spectra, are carried out correctly and precisely. As a rule of thumb, there are a number of key considerations, as concisely given in [46], and these are:

1. Clearly understand how the spectra are measured, i.e., knowing the instrumental resolution, sampling method used, and any spectral preprocessing, i.e., subtraction, smoothing, baseline correction, and other processing algorithms applied like spectral derivation. All of these factors will affect the appearance of the spectra.

2. When using wavenumber (preferable for most spectroscopist), the spectra are being plotted with highest wavenumber at the most left of the graph and followed by smaller wavenumber running to the right.

3. The peaks of the spectra should be on scale, i.e., between absorbance units of 0 and 2 (or transmittance units of 10–90%); otherwise, it is considered to be off scale

4. Measured spectra should be in a "good quality," i.e., low noise, little or no baseline offset, a flat baseline, peaks on scale, and no spectral artifacts (i.e., the unwanted spectra that are usually contributed during the sample measurement). The use of background correction, sealing and desiccating the measuring instrument, as well as purging measurement area with dry N_2 gases are typical efforts used to minimize occurrence of spectral artifacts.

5. Identify any possible spectral artifacts before assigning other peaks.

6. Use other supporting knowledge about the examined sample, i.e., how the samples being prepared, its physical properties and appearances, or when possible, analysis results using other measurement techniques such as Raman, UV–Vis, or nuclear magnetic resonance (NMR) in order to make further accurate interpretation tasks easier.

7. Identify the positions of common functional groups or chemical bonds that might be picturized in the acquired spectra. Spectral absorption ranges from common functional groups given in **Table 1**.

8. Assign the most intense bands first, followed by ones with lesser intensities. One does not need to identify all peaks in the spectra, but most importantly, it is the spectra that are most important for the goal of analysis.

9. In the case that the changes in the intended observed effect are not clearly seen in an overlapped spectral band, deconvoluting the spectra into its constructing peaks would be helpful to discern the observed effect.

10. In several cases, the pattern changes in observed spectral bands are not clearly "readable" from the spectra itself. Thus, an additional analysis tool is then

Position of peaks		**Functional groups or Chemical Bond**
[cm^{-1}]	**[nm]**	
3300–3500	2857.1–3030.3	**N** – **H** stretch
3200–3400	2941.2–3125.0	**O** – **H** stretch
2850–3100	3225.8–3508.8	**C** – **H** stretch
2100–2260	4424.8–4761.9	**C** = **N** strech **C** ≡ **C** strecth
1650–1800	5555.6–6060.6	**C** = **O**stretch
500–1000	10,000–20,000	**C** = **C** stretch **C** – **H** wagging aromatic ring *fingerprint* region

Table 1.
Positions of some common functional groups or chemical bonds.

required to make the interpretation of the pattern of spectral changes easier to extract. Chemometrics analysis can be beneficial to discern these spectral patterns.

5.5 Chemometric analysis

Infrared Spectroscopy provides spectral data that contains extremely large amounts of information, as a fingerprinting tool to extract several important properties of the measured sample [47]. Interpretation of infrared spectra is a complex process, since the spectra show peaks and vibrations of molecular bonds in the measured sample, as well as their combinations and overtones. Chemometrics is thus required to interpret this data [48].

Chemometrics, according to the International Chemometrics Society, is the science of relating measurements made on a chemical system or process to the state of the system using mathematical or statistical methods. Since its first introduction in the 1960s, chemometrics has become dispersed in many areas [49]. It is also widely used as an essential tool in modern analytical Vibrational-Spectroscopy-based instruments to efficiently extract the maximum useful information from spectra in a relatively short time.

The most commonly used chemometric methods in IR spectral analysis can be classified into the following groups [50]: the processing techniques (e.g., normalizations, baseline corrections, centering, derivatives, multiplicative corrections, and required combinations of them) to enhance the information contained in the spectra; the classification methods (either supervised or unsupervised), which link any a priori knowledge about the samples to be classified; and the regression methods, which link the spectra to quantifiable properties of the samples, mainly those based on multivariate regression through principal component analysis (PCA) and partial least squares regression (PLSR) and Discriminant Analysis (DA) approaches. In PCA, the correlated response variables in spectral data are transformed into non-correlated variables known as principal components (PC's) [51]. Meanwhile, the PLSR is used to reduce the predicting variables to a smaller set of predictors, which are then used to perform a regression [52].

6. Detecting adulterants in food and traditional Indonesian herbal medicine using infrared spectroscopy

In this section, we will devote our discussion on how to recognize the adulteration based on the IR Spectra acquired. The use of necessary preprocessing techniques will be also addressed to make the changes in spectral pattern more easily readable. Cases of adulterants discussed in this section are as previously described in Sections 2-4.

6.1 Over-dilution with water or other solvents

Dilution is the process of adding additional water or other solvents to a product in order to increase total volume while decreasing the concentration of active ingredients. This procedure is often used to adulterate milk, juices, and other liquid food or medicinal products. In order to recognize the dilution, one should observe and analyze peaks of water molecule's spectral signature embedded in the spectrum of the investigated samples. Absorption spectrum of water molecule is given in **Figure 8**.

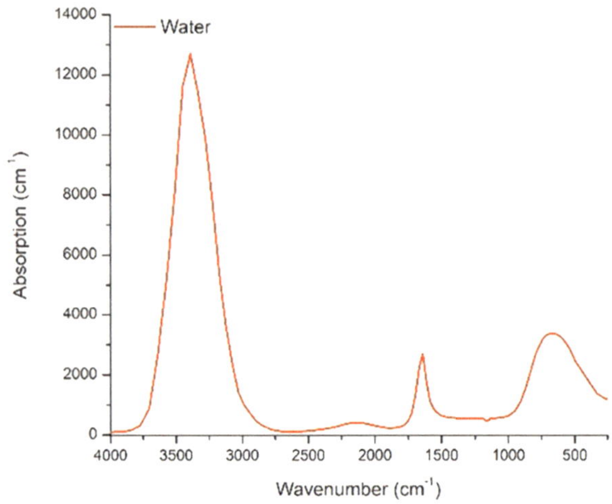

Figure 8.
Water molecule absorption spectrum (reproduced from [53]).

Breast milk is regarded as the optimum exclusive source of nutrition for the first 6 months of baby's life and may remain part of the healthy infant diet for the first 2 years of life and beyond. The WHO and UNICEF recommend that a baby initiate breastfeeding within the first hour of birth and be exclusively breastfed for the first 6 months of life [54]. Unfortunately, not all *postpartum* mothers can provide sufficient breastfeeding for their babies; therefore, in some countries there are noble initiatives to establish a human milk banking [55]. These milk banking systems collect breast milk from donor mothers, who have more than their babies need, process and redistribute the donated milk to the needed babies, i.e., the premature and sick babies whose mothers do not have enough milk for them.

Sadly, these noble initiatives were disgraced by irresponsible acts to adulterate the milk. In this context, efforts to develop a screening system to differentiate between real BM and its adulterated ones are necessary to guarantee that the needed babies receive the required nutrition available in BM. Infrared Spectroscopy proved to be beneficial as a screening technique for this problem.

One interesting case to discuss here is the detection of adulterated milk as described in [56]. This paper analyzed the difference among Human Breastmilk (BM) and related adulterated with water (W) and cow milk (CM), semi-skimmed cow milk (SSCM), and skimmed cow milk (SCM). The recorded FTIR spectra of the comparison among original BM and adulterated variants are given in **Figure 9**. From the spectra we can observe the prominent absorption peaks of water, as in **Figure 8**, i.e., in regions 3500–3000 cm^{-1} and 1730–1600 cm^{-1} due to H_2O stretching and H-O-H bending vibration. These peaks seem to be overlapped with other absorption bands of other molecules in milk, make them difficult to clearly discern visually. The other absorption bands observed in the ranges 1630–1680 cm^{-1} and 1510–1570 cm^{-1} are due to C=O stretching or N-H and C-H bending vibration of the milk proteins, and the bands of 2920, 2850, and 1743 cm^{-1} may be due to the antisymmetric and symmetric stretching of CH_2 and C=O groups from the fatty milk components, respectively [57].

Meanwhile the absorption bands in the ranges 3200–3700, 1030–1200, and 900–450 cm^{-1} have been associated with carbohydrate structures. That is why other processing technique such as derivative, PCA, and PLSR are needed to get better

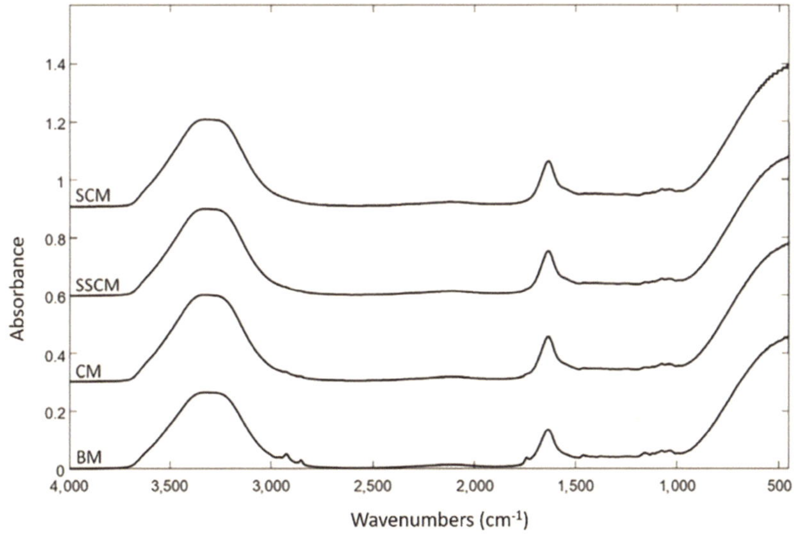

Figure 9.
FTIR spectra comparison of pure BM, CM, SCM, and SSCM.

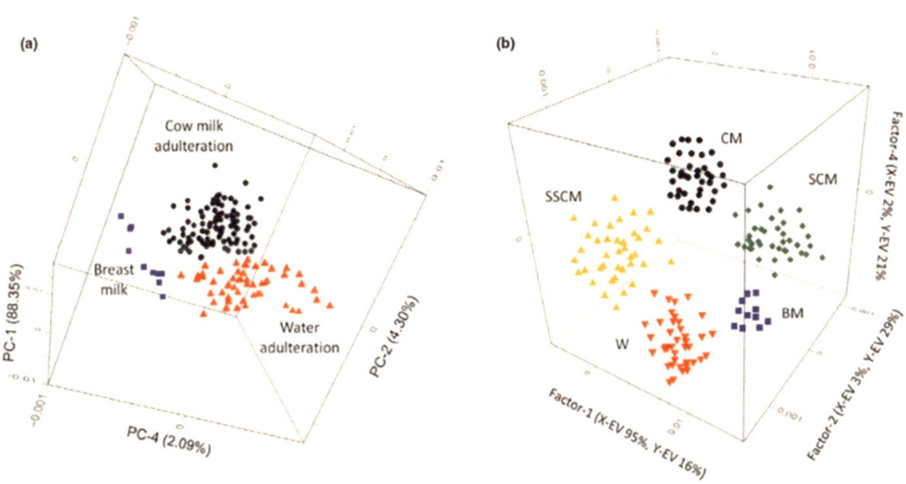

Figure 10.
(a) 3D score plot by PCA and (b) by PLSR-DA [56].

insight into the separation between real and adulterated milk, as can be seen in **Figure 10**.

6.2 Boosting the nitrogen content in milk

In the year 2008, the world was shocked by the deadly milk scandal in China, where around 300,000 children were poisoned by melamine, a chemical that is usually used to make plastic that was added to powdered milk to increase nitrogen content in milk. Melamine consumption may result in reproductive damage, bladder

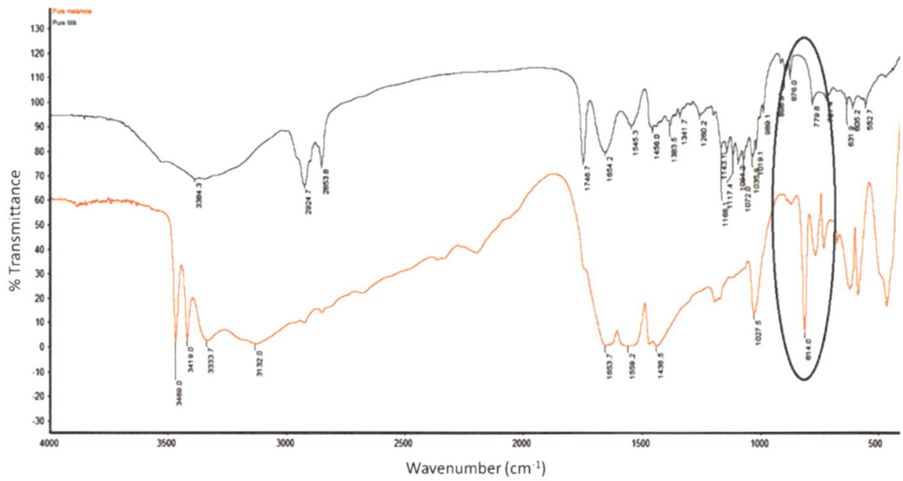

Figure 11.
FTIR transmittance spectra of milk and melamine [58].

or kidney stones, and bladder cancer. Infrared Spectroscopy was beneficial as a rapid and accurate technique in screening the availability of melamine in milk. There is a distinct absorption band of melamine, i.e., characterized by the out-of-plane bending of the 1,3,5-s-triazine ring of melamine at 814 cm^{-1}, which is absent in pure milk. This spectral difference can be depicted in **Figure 11**. Using region of 851.62–798.39 cm^{-1} as calibration, Jawaid et al. [58] obtained a good PLS model for very low detection range up to 1%, i.e., with a good correlation coefficient of $R^2 = 0.999$ and accuracy and high precision (i.e., 0.12–1.10% error and relative standard deviation of 1.38–2.07%, respectively), and with limit of detection (LOD) can achieve 1 ppm.

6.3 Artificial sweeteners

Sugar (sucrose) additions in food and drinks serve several functions; among other is changing the flavor profile to increase appetite and likelihood of consumption. It is also used to provide a sweet taste to lessen the bitter and sour taste of some ingredients in some medicinal or traditional herbal drinks (Jamu). Historically, the use of artificial sweeteners has been used a long time ago [59] as an enhancer of the sweet taste in food/drink products or as a substitute for sugar for those with a problem of sugar consumption. Unfortunately, it has been found in recent decades that consumption of these kinds of sweeteners may induce metabolic disorders [60, 61].

IR spectra from sucrose and other common artificial sweeteners can be seen in **Figure 12**.

There are many overlapped absorption peaks observed, even though there are some specific peaks observed that might characterize each of these sweeteners, as well as differ from sucrose, which can be used as a rapid non-destructive tool for sweetener identification. As in the case of milk, the use of chemometric analysis can accurately classify the distinct signatures of sweeteners.

Interesting work is reported by Wang et al. [63] that builds and tests PLS model to identify differentiate single and double mixtures of artificial sweeteners. Mixture of sweeteners usually done to reduce bitter off-taste following its consumption [64]. The

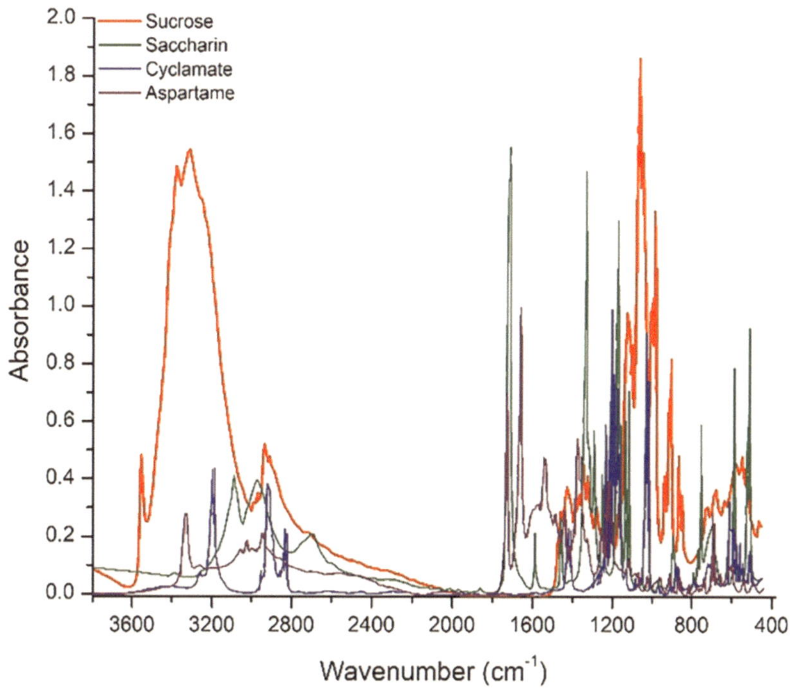

Figure 12.
Absorbance spectra of several common artificial sweeteners, reproduced from [62].

developed model may have prediction performance with R^2 ranging from 0.9981 to 0.9996 for a single type of sweetener, and from 0.9397 to 0.9998 for a blend of two types of sweeteners. Blends of more sweetener types may produce lower fitting degree of prediction and measured value, i.e., 0.7648–0.9997, 0.7292–0.9994, 0.6617–0.9968, for three, four, and five sweetener mixtures, respectively.

6.4 Adulterated Jamu with active pharmaceutical ingredients (APIs)

Some active pharmaceutical substances found to be added in the illegally marketed herbal medicine in Indonesia. Paracetamol, phenylbutazone, and dexamethasone are among chemical substances that are usually added to Jamu, in order to provide pain relieving effects and comfortable sensations to the consumers. Unfortunately, many consumers do not realize the harmful side effects that can be induced by consuming these adulterated Jamu [65].

In order to identify the addition of APIs in Jamu, one should know how the IR spectrum of each API differs specifically. Different absorbance IR spectra from paracetamol, phenylbutazone, and dexamethasone are depicted in **Figure 13**. It is observed that some distinct unique absorption peaks can be used for recognition.

There were also evidence that in some arthritic and rheumatic herbal medicine (jamu pegal linu) was adulterated with the Piroxicam, i.e., a type of nonsteroidal anti-inflammatory drug (NSAID) used to relieve the symptoms of painful inflammatory conditions such as arthritis and other similar chemical such as prednisone and naproxen [13, 65]. **Figure 14** provides IR transmission spectra of piroxicam, and its typical absorption bands are given in **Table 2**.

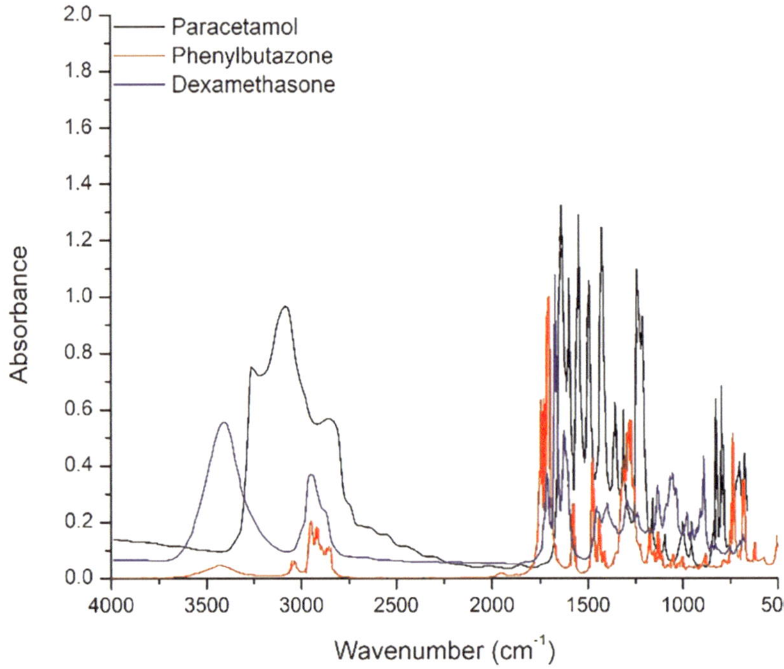

Figure 13.
Absorbance spectra of paracetamol, phenylbutazone, and dexamethasone, reproduced from [62].

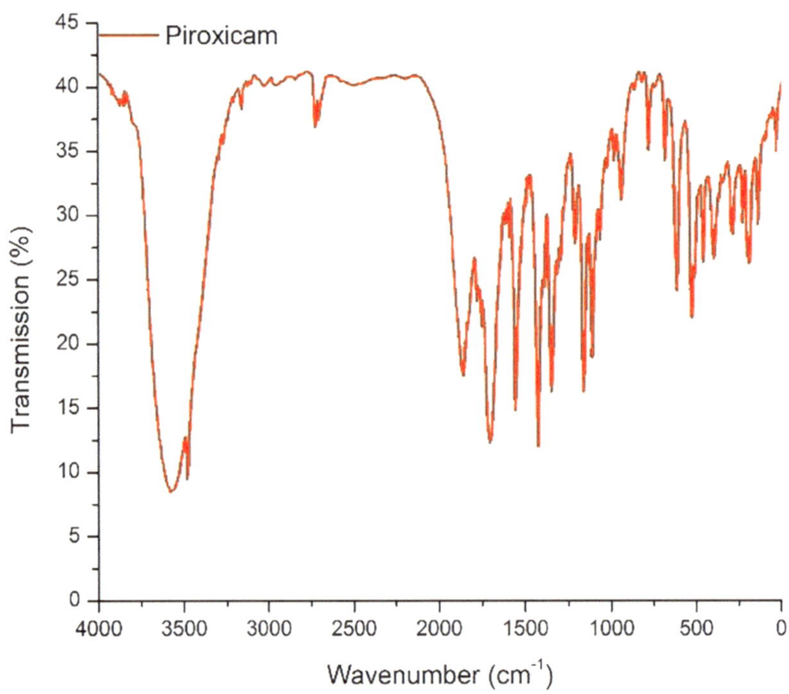

Figure 14.
IR transmission spectra of piroxicam.

Frequency (cm^{-1})	Band Assignments
3854	OH stretching
3338	NH stretching
2365	CH stretching
1629	Amide **C**=**O** stretching
1527	Aromatic **C**=**C** stretching
1435	**N**—**H** bending
1181	**C**—**S** stretching
938, 830	**C**—**H** bending

Table 2.
Infrared absorption bands of piroxicam.

Readers who are interested in deepening their understanding on how IR Spectroscopy can be beneficially used to authenticate herbal products can refer to these interesting review papers [66–68].

7. Summary and future outlooks

In this chapter, evidence of food fraud and types of adulteration in food and medicinal products has been described. Indonesian traditional herbal products, as traditional medication ways that are widely practiced among societies residing in the Indonesian archipelagos, are also prone to irresponsible adulteration practices.

Infrared Spectroscopy has proved to be a sensitive and rapid identification tool for recognizing these adulterants. Fundamental principles for generating, measuring, and analyzing the specific spectrum of adulterant molecules have also been described. Infrared spectra contain extremely large amounts of information as a fingerprinting tool to extract several important properties of the measured sample, and the elaboration of chemometric analysis makes the analysis of the investigated samples more clear and easier to handle.

Grasping better insight into the responsible spectral bands that are strongly correlated with the observed variables would be beneficial in developing a customized portable measurement system [69]. Such a system might be helpful in terms of technology to keep people safe from adulterated food and medicinal products.

Author details

Aulia M.T. Nasution[1]* and Hery Suyanto[2]

1 Faculty of Industrial Technology and System Engineering, Institut Teknologi Sepuluh Nopember (ITS), Photonics Engineering Laboratory, Department of Engineering Physics, Surabaya, Indonesia

2 Faculty of Mathematics and Natural Sciences, Integrated Basic Sciences Laboratory, Udayana University, Bukit Jimbaran, Kuta Selatan, Badung-Bali, Indonesia

*Address all correspondence to: anasution@ep.its.ac.id

References

[1] Bansal S, Singh A, Mangal M, Mangal AK, Kumar S. Food adulteration: Sources, health risks, and detection methods. Critical Reviews in Food Science and Nutrition. 2017;**57**(6): 1174-1189. DOI: 10.1080/10408398.2014.967834

[2] Hutt PB. The basis and purpose of government regulation of adulteration and misbranding of food. Food, Drug, Cosmetic Law Journal. 1978;**33**(10): 505-540

[3] Banati D. Adulteration of foodstuffs: From misleading to poisoning. In: Elmadfa I, editor. Diet Diversification and Health Promotion. Vol. 57. Basel, Switzerland: S. Karger AG; 2005. pp. 124-134

[4] Tan CH, Kong I, Irfan U, Solihin MI, Pui LP. Edible oils adulteration: A review on regulatory compliance and its detection technologies. Journal of Oleo Science. 2021;**70**(10):1343-1336. DOI: 10.5650/jos.ess21109

[5] Sahoo N, Manchikanti P, Dey S. Herbal drugs: Standards and regulation. Fitoterapia. 2010;**81**(6): 462-471. DOI: 10.1016/j.fitote.2010.02.001

[6] FAO. Food Security, FAO Policy Brief. 2006. pp. 1-4

[7] Spink J, Moyer DC. Defining the public health threat of food fraud. Journal of Food Science. 2011;**76**(9): R157-R163. DOI: 10.1111/j.1750-3841.2011.02417.x

[8] Graham MR et al. Counterfeiting in performance- and image-enhancing drugs. Drug Testing and Analysis. 2009;**1**(3):135-142. DOI: 10.1002/dta.30

[9] Cole C, Jones L, McVeigh J, Kicman A, Syed Q, Bellis M. Adulterants in illicit drugs: A review of empirical evidence. Drug Testing and Analysis. 2011;**3**(2):89-96. DOI: 10.1002/dta.220

[10] Calahan J, Howard D, Almalki AJ, Gupta MP, Calderón AI. Chemical adulterants in herbal medicinal products: A review. Planta Medica. 2016;**82**(6): 505-515. DOI: 10.1055/s-0042-103495

[11] El-Jardali F et al. Interventions to combat or prevent drug counterfeiting: A systematic review. BMJ Open. 2015;**5**: 11. DOI: 10.1136/bmjopen-2014-006290

[12] Yéléhé-Okouma M et al. Drug adulteration of sexual enhancement supplements: A worldwide insidious public health threat. Fundamental & Clinical Pharmacology. 2021;**35**(5): 792-807. DOI: 10.1111/fcp.12653

[13] Syahfitr SA, Asra DR. Analysis of medicinal chemicals contained on Jamu: A review. Asian Journal of Pharmaceutical Research and Development. 2021;**9**(2):33-46. DOI: 10.22270/ajprd.v9i2.931

[14] Richardson PIC, Muhamadali H, Ellis DI, Goodacre R. Rapid quantification of the adulteration of fresh coconut water by dilution and sugars using Raman spectroscopy and chemometrics. Food Chemistry. 2019; **272**:157-164. DOI: 10.1016/j.foodchem.2018.08.038

[15] Moore JC, Spink J, Lipp M. Development and application of a database of food ingredient fraud and economically motivated adulteration from 1980 to 2010. Journal of Food Science. 2012;**77**(4):R118-R126. DOI: 10.1111/j.1750-3841.2012.02657.x

[16] Azad T, Ahmed S. Common milk adulteration and their detection techniques. International Journal of Food Contamination. 2016;**3**(1). DOI: 10.1186/s40550-016-0045-3

[17] Yang R, Huang W, Zhang L, Thomas M, Pei X. Milk adulteration with melamine in China: Crisis and response. Quality Assurance and Safety of Crops & Foods. 2009;**1**(2):111-116. DOI: 10.1111/j.1757-837X.2009.00018.x

[18] Cohen PA. The FDA and adulterated supplements-dereliction of duty. JAMA Network Open. 2018;**1**(6):e183329. DOI: 10.1001/jamanetworkopen.2018.3329

[19] Tucker J, Fischer T, Upjohn L, Mazzera D, Kumar M. Unapproved pharmaceutical ingredients included in dietary supplements associated with US food and drug administration warnings. JAMA Network Open. 2018;**1**(6):e183337. DOI: 10.1001/jamanetworkopen.2018.3337

[20] Yun J, Choi J, Jo CH, Kwon K. Detection of synthetic anti-obesity drugs, designer analogues and weight-loss ingredients as adulterants in slimming foods from 2015 to 2017. Journal of Chromatography and Separation Techniques. 2018;**09**(01). DOI: 10.4172/2157-7064.1000396

[21] Cohen PA, Avula B, Wang YH, Zakharevich I, Khan I. Five unapproved drugs found in cognitive enhancement supplements. Neurology Clinical Practice. 2021;**11**(3):e303-e307. DOI: 10.1212/cpj.0000000000000960

[22] De Bruyn M et al. Borneo and Indochina are major evolutionary hotspots for Southeast Asian biodiversity. Systematic Biology. 2014;**63**(6):879-901. DOI: 10.1093/sysbio/syu047

[23] Lohman DJ et al. Biogeography of the Indo-Australian Archipelago. Annual Review of Ecology, Evolution, and Systematics. 2011;**42**:205-226. DOI: 10.1146/annurev-ecolsys-102710-145001

[24] Elfahmi, Woerdenbag HJ, Kayser O. Jamu: Indonesian traditional herbal medicine towards rational phytopharmacological use. Journal of Herbal Medicine. 2014;**4**(2):51-73. DOI: 10.1016/j.hermed.2014.01.002

[25] Javanologi PUI. Traces of Jamu as herbal medicine in history. 2021. Available from: https://javanologi.uns.ac.id/en/2021/09/02/jejak-jamu-dalam-goresan-sejarah-copy/

[26] Beers S-J. Jamu: The Ancient Art of Herbal Healing. Tokyo: Tuttle Publishing; 2001

[27] Lubis MF, Zaitun Hasibuan PA, Syahputra H, Surbakti C, Astyka R. Saurauia vulcani (Korth.) as herbal medicine potential from North Sumatera, Indonesia: A literature review. Heliyon. 2022;**8**(4):4-9. DOI: 10.1016/j.heliyon.2022.e09249

[28] Agustina ZA, Fitrianti Y. Utilization of Jamu in Puerperal Mother in Sumatera and Java Island (Literature Review of Health Ethnographic Research 2012-2016). Indonesian Journal of Public Health. 2020;**15**(1):93. DOI: 10.20473/ijph.v15i1.2020.93-102

[29] Diliarosta S, Prima Sari M, Ramadhani R, Efendi A. Ethnomedicine study on medicinal plants used by communities in West Sumatera, Indonesia. In: El-Shemy H, editor. Natural Medicinal Plants. London: InTechOpen; 2022. p. 11

[30] Jannah M, Afifah N, Hariri MR, Rahmawati A, Wulansari TYI. Study of lichen (Usnea spp.) as a traditional medicine in Bogor, West Java. Berkala Penelitian Hayati. 2020;**26**(1):32-38. DOI: 10.23869/bphjbr.26.1.20206

[31] Iskandar BS, Iskandar J, Partasasmita R, Irawan B. Various medicinal plants traded in the village market of Karangwangi Village, Southern Cianjur, West Java, Indonesia. Biodiversitas. 2020;**21**(9):4440-4456. DOI: 10.13057/biodiv/d210963

[32] Riptanti EW, Qonita A, Fajarningsih RU. Potentials of sustainable development of medicinal plants in Wonogiri regency of Central Java province of Indonesia. Bulgarian Journal of Agricultural Science. 2018; **24**(5):742-749

[33] Fathir A, Haikal M, Wahyudi D. Ethnobotanical study of medicinal plants used for maintaining stamina in madura ethnic, East Java, Indonesia. Biodiversitas. 2021;**22**(1):386-392. DOI: 10.13057/biodiv/d220147

[34] Diba F, Yusro F, Mariani Y, Ohtani K. Inventory and biodiversity of medicinal plants from tropical rain forest based on traditional knowledge by ethnic dayaknese communities in west kalimantan indonesia. Kuroshio Science. 2013;7(1):75-80

[35] Falah F, Hadiwibowo N. Species identification of traditional medicine plants for women's health in East Kalimantan: Lesson learned from local wisdom. Indonesian Journal of Forestry Research. 2017;**4**(1):49-68. DOI: 10.20886/ijfr.2017.4.1.49-68

[36] Maharani R, Fernandes A. The potential of herbal medicine from Kalimantan, Indonesia, to stimulate human immunity during the COVID-19 pandemic: A brief overview. Annals of Phytomedicine : An International Journal. 2021;**10**(Sp-Issue1 COVID-19): S4-S12. DOI: 10.21276/ap.covid19.2021.10.1.11

[37] Pandiangan D, Silalahi M, Dapas F, Kandou F. Diversity of medicinal plants and their uses by the Sanger tribe of Sangihe Islands, North Sulawesi, Indonesia. Biodiversitas. 2019; **20**(3):621-631. DOI: 10.13057/biodiv/d200301

[38] Rahmawati N, Mustofa FI, Haryanti S. Diversity of medicinal plants utilized by to manui ethnic of central Sulawesi, Indonesia. Biodiversitas. 2020; **21**(1):375-392. DOI: 10.13057/biodiv/d210145

[39] Mustofa FI, Rahmawati N, Aminullah. Medicinal plants and practices of rongkong traditional healers in South Sulawesi, Indonesia. Biodiversitas. 2020;**21**(2):642–651. DOI: 10.13057/biodiv/d210229

[40] Dirgantara S, Tanjung RHR, Nurlatifah R, Meiyanto E. Cytotoxic activity of selected medicinal plants from Papua, Indonesia. In: Proceedings of BROMO Conference (BROMO 2018), no. Bromo. 2018. pp. 155–158. DOI: 10.5220/0008359001550158

[41] Budiarti M et al. The use of antimalarial plants as traditional treatment in Papua Island, Indonesia. Heliyon. 2020;**6**(12):e05562. DOI: 10.1016/j.heliyon.2020.e05562

[42] Fatimah S, Rahayu M, Rinding ALT. Saccharin analysis in Jamu Kunyit Asam (tumeric acid) sold in Malioboro and in Beringharjo traditional market Yogyakarta. Biomedika. 2017;**10**(1):

30-35. DOI: 10.31001/biomedika.v10i1.222

[43] Halonen L. Local mode vibrations in polyatomic molecules. In: Advances in Chemical Physics. Vol. 104. New York: John Wiley & Sons, Inc; 1998. pp. 41-179

[44] Kellman ME, Tyng V. The dance of molecules: New dynamical perspectives on highly excited molecular vibrations. Accounts of Chemical Research. 2007; **40**(4):243-250. DOI: 10.1021/ar000153r

[45] Grim WM III, Fateley WG, Graselli JG. Introduction to dispersive and interferometric infrared spectroscopy. In: Theophanides T, editor. Fourier Transform Infrared Spectroscopy. Québec, Canada: D. Reidel Publishing Company; 1984. pp. 25-42

[46] Smith B. Infrared Spectral Interpretation: A systematic approach. Boca Raton, FL: CRC Press; 1999

[47] dos Santos CAT, Páscoa RNMJ, Lopes JA. A review on the application of vibrational spectroscopy in the wine industry: From soil to bottle. Trends in Analytical Chemistry. 2017;**88**:100-118. DOI: 10.1016/j.trac.2016.12.012

[48] Varmuza K, Filzmoser P. Introduction to Multivariate Statistical Analysis in Chemometrics. Boca Raton, FL: CRC Press; 2009

[49] Brereton RG. A short history of chemometrics: A personal view. Journal of Chemometrics. 2014;**28**(10):749-760. DOI: 10.1002/cem.2633

[50] Moros J, Garrigues S, de la Guardia M. Vibrational spectroscopy provides a green tool for multi-component analysis. Trends in Analytical Chemistry. 2010; **29**(7):578-591. DOI: 10.1016/j.trac.2009.12.012

[51] Sato T. Application of principal-component analysis on near-infrared spectroscopic data of vegetable oils for their classification. Journal of the American Oil Chemists' Society. 1994; **71**(3):293-298. DOI: 10.1007/BF02638055

[52] Luinge HJ, van der Maas JH, Visser T. Partial least squares regression as a multivariate tool for the interpretation of infrared spectra. Chemometrics and Intelligent Laboratory Systems. 1995;**28**(1):129-138. DOI: 10.1016/0169-7439(95)80045-B

[53] Hale GM, Querry MR. Optical Constants of Water in the 200-nm to 200-μm Wavelength Region. Applied Optics. 1973;**12**(3):555. DOI: 10.1364/ao.12.000555

[54] WHO. Breastfeeding. Available from: https://www.who.int/health-topics/breastfeeding#tab=tab_1.

[55] Haiden N, Ziegler EE. Human Milk Banking. Annals of Nutrition & Metabolism. 2017;**69**(2):8-15. DOI: 10.1159/000452821

[56] De Luca M, Ioele G, Spatari C, Caruso L, Galasso MP, Ragno G. Evaluation of human breastmilk adulteration by combining Fourier transform infrared spectroscopy and partial least square modeling. Food Science & Nutrition. 2019;**7**(6): 2194-2201. DOI: 10.1002/fsn3.1067

[57] Carbonaro M, Nucara A. Secondary structure of food proteins by Fourier transform spectroscopy in the mid-infrared region. Amino Acids. 2010; **38**(3):679-690. DOI: 10.1007/s00726-009-0274-3

[58] Jawaid S, Talpur FN, Afridi HI, Nizamani SM, Khaskheli AA, Naz S.

Quick determination of melamine in infant powder and liquid milk by Fourier transform infrared spectroscopy. Analytical Methods. 2014;**6**(14): 5269-5273. DOI: 10.1039/c4ay00558a

[59] Eggleston G. In: Orna MV, Eggleston G, Bopp AF, editors. History of Sugar and Sweeteners, in Chemistry's Role in Food Production and Sustainability: Past and Present. Washington, D.C.: American Chemical Society; 2019

[60] Swithers SE. Artificial sweeteners produce the counterintuitive effect of inducing metabolic derangements. Trends in Endocrinology and Metabolism. 2013;**24**(9):431-441. DOI: 10.1016/j.tem.2013.05.005

[61] Pepino MY. Metabolic effects of non-nutritive sweeteners. Physiology & Behavior. 2015;**152**:450-455. DOI: 10.1016/j.physbeh.2015.06.024

[62] The National Institute of Standards and Technology (NIST). NIST Standard Reference Database 69: NIST Chemistry WebBook. Available from: https://webbook.nist.gov/chemistry/

[63] Wang YT et al. FTIR spectroscopy coupled with machine learning approaches as a rapid tool for identification and quantification of artificial sweeteners. Food Chemistry. 2020;**303**:1-11. DOI: 10.1016/j.foodchem.2019.125404

[64] Behrens M, Blank K, Meyerhof W. Blends of non-caloric sweeteners saccharin and cyclamate show reduced off-taste due to tas2r bitter receptor inhibition. Cell Chemical Biology. 2017; **24**(10):1199-1204.e2. DOI: 10.1016/j.chembiol.2017.08.004

[65] Suparmi S, Wahidin D, Rietjens IMCM. Risk characterisation of constituents present in jamu to promote its safe use. Critical Reviews in Toxicology. 2021;**51**(2):183-191. DOI: 10.1080/10408444.2021.1912708

[66] Rohman A, Windarsih A, Motalib Hossain MA, Johan MR, Ali ME, Nurulhidayah AF. Application of near- and mid-infrared spectroscopy combined with chemometrics for discrimination and authentication of herbal products: A review. Journal of Applied Pharmaceutical Science. 2019; **9**(3):137-147. DOI: 10.7324/JAPS.2018.8801

[67] Bunaciu AA, Aboul-Enein HY. Adulterated drug analysis using FTIR spectroscopy. Applied Spectroscopy Reviews. 2021;**56**(5):423-437. DOI: 10.1080/05704928.2020.1811717

[68] Fatmarahmi DC, Susidarti RA, Swasono RT, Rohman A. A development method of FTIR spectroscopy coupled with chemometrics for detection of synthetic drug adulterants of herbal products in quaternary mixture. Journal of Applied Pharmaceutical Science. 2022; **12**(3):191-201. DOI: 10.7324/JAPS.2022.120320

[69] Müller-Maatsch J, van Ruth SM. Handheld devices for food authentication and their applications: A review. Food. 2021;**10**(12):1-22. DOI: 10.3390/foods10122901

Chapter 10

Application of Raman Spectroscopy for Dental Enamel Surface Characterization

Cecilia Carlota Barrera-Ortega,
America Rosalba Vazquez Olmos, Roberto Isaac Sato Berrú
and Pineda Dominguez Karla Itzel

Abstract

Dental enamel is the most complex and highly mineralized human body tissue, containing more than 95% of carbonated hydroxyapatite and less than 1% of organic matter. Current diagnostic methods for enamel caries detection are unable to detect incipient caries lesions. Many papers determine the re-mineralizing effect using many fluorinated compounds and different demineralizing solutions to test physical characterizations such as microhardness, roughness, wettability, among others, but there is not much information about the use of Raman Spectroscopy. Raman Spectroscopy is an efficient technique of chemical characterization to identify functional groups (phosphate-hydroxyl groups) found in the hydroxyapatite formula, which helps identify the level of mineralization on dental enamel surface. Raman spectroscopy is applicable to any state of aggregation of the material, indicated for biological samples. Given the minimum bandwidth of a laser source, as with all spectroscopic techniques that use a laser source, a small sample is sufficient, which makes it an important technique in the analysis of reactive products with very low yield. Raman spectroscopy can be used to obtain the main functional groups in order to determine the remineralization of dental enamel; these results are highly valuable as they can help us make the best decisions on dental treatments.

Keywords: dental enamel, remineralization, demineralization, Raman spectroscopy, hydroxyapatite

1. Introduction

In mineralized biological systems, it has been found in the literature that there are different ways to administer fluoride, like the application of varnishes, tablets, and different dental pastes with different concentrations of fluoride that participate in an important way in the mineralization mechanisms of the fundamental unit of enamel (hydroxyapatite prisms), modifying the chemical composition, and increasing the resistance to dissolution in an acidic environment. To inhibit the formation of

IntechOpen

demineralized lesions and the progression to carious lesions, fluorinated compounds are currently applied to the external surface of the enamel. However, the lack of information on the different vehicles or concentrations of fluorinated compounds, as well as the extent of the effect on enamel, leads to the use of these compounds being exaggerated and at times ineffective in preventive dentistry.

Dental enamel is the outer covering of dental crowns, also known as adamantine tissue, and it is currently defined as a nanocomposite bioceramic of epithelial origin, which protects the tooth from chemical and physical aggressions, which has been considered the most mineralized and hard tissue of the organism because it is structurally constituted by millions of highly mineralized prisms that run through its entire thickness, from the amelo-dentin junction to the external or free surface in contact with the oral environment [1].

Its specific function is to form a resistant cover for the teeth that will make them suitable for chewing. In charge of covering and protecting the dentin-pulp complex from chemical and physical aggressions, it lacks vascularization and innervation, which prevents its own remodeling or repair [2].

Embryologically it is widely known that dental enamel is of ectodermal origin, and the formation of this dental structure occurs by cellular events collectively called amelogenesis and biochemical events that are called biomineralization.

The chemical composition of dental enamel is made up of 95% inorganic matter, organic matter 4%, and water 1% [3].

2. Raman and infrared spectroscopy

The symmetric vibrational properties of the molecules are used in systematic ways to interpret the infrared spectrum, as they can be used to predict the transitional vibrations allowed, practically only using the table of characters of the punctual group the molecule belongs to. Raman spectral analysis is often compared with the well-known infrared absorption (IR) spectroscopy. While the two techniques are similar, they work in distinct ways and measure different things. The IR technique measures light absorption by specific molecules, while the Raman technique measures Raman emission from molecules under monochromatic laser irradiation. The difference between the light signals and Raman emission corresponds to the vibrational frequencies of these molecules. The two techniques by themselves are great for obtaining important information from samples but the two can be used in combination to measure vibrational bands unique to each technique; that is why the IR and Raman techniques are often regarded to be complementary.

As mentioned earlier, IR spectroscopy is an absorption technique and the measurements are determined by changes in vibrational frequency, whereas Raman spectroscopy uses a scattering method to obtain data from changes in the polarizability tensor. These differences affect both the method of obtaining data from samples and the parameters that are necessary for calibration curves [4–6].

Raman spectroscopy can be applied to any state of aggregation: solids, liquids, or gases. In liquid dissolutions, this technique presents advantages over infrared spectroscopy, as only waves of longitude of the visible region of the spectrum are implied, so only the cells and glass optics are precise. Also, water produces very weak Raman signals, which will not tangle the spectrum. These advantages make Raman spectroscopy specially indicated for biological samples. Given the minimum bandwidth of a source laser, a small quantity of the sample is sufficient, which makes this an important technique to analyze the reactions of products with a low yield. Other advantages are derived from the fact that as visible radiation is being employed, and overheating of the samples is notably reduced. The most notable difficulty of Raman spectroscopy is the fluorescence that the enamel emits after the application of different fluoride components such as gels, varnishes, or toothpastes.

2.1 Other techniques used for characterizing dental enamel

The mineral content of dental enamel confers itself mechanical, physical, chemical, and biological properties. As it is the most exposed tissue and therefore the most susceptible to demineralization by acidic agents, different techniques have been used throughout time to better know the dental enamel. The first observations were made using scanning electronic microscopy (SEM) and semiquantitative elemental percentage analysis, and to determine the roughness of the dental enamel, atomic force microscopy, and contact and now optical profilometers have been used. To establish the superficial energy of the enamel, wettability is measured by calculating the contact angle, and this property gives us information on how hydrophobic or hydrophilic the dental enamel is. Other techniques used are the nano- and micro-hardness tests, which have been used as physical characterization to assess quantitative demineralization or remineralization of the dental enamel, and finally, Raman spectroscopy is the gold standard to determine the presence or absence of the phosphate-hydroxyl functional groups in dental enamel.

2.1.1 Strength

The strength is a mechanical property of any material, consistent with the difficulty that exists to scratch or create mark on a surface by means of a penetrating point. Strength is measured using a strength tester to rehearse penetration. Depending on the type of point used and the range of loads applied, different scales exist, adequate for different ranges of strength [7].

The Vickers hardness test (HVN [hardness Vickers number]) consists in marking the testing material with a diamond indenter that has the form of a pyramid with a square base and angle of 136° in between opposing faces, put through a load of 1 to 100 kg/f.

Microhardness tests are used a lot and have an important application in dentistry. A microhardness test can evaluate the level of mineralization of a dental substrate. A specific force applied during a specific time and distance provides important data. It has the capacity of remineralizing the enamel and the dentine after different treatments, like it happens in unbalanced situations of demineralization and remineralization [8].

The roughness is a property of a material, speaking of their surface. The superficial roughness is a set of irregularities on the real surface, defined conventionally in a section where deformities and undulations have been eliminated. The appearance of the surface of a piece depends primarily on the material with what the piece was made of and its conformation process. To currently measure this property, a profilometer is used, which is an optic device that has the advantage of a no contact exploration, therefore avoiding deformities and possible harm of soft surfaces. They can also explore surfaces that are not accessible to mechanical devices, measuring through a transparent layer and measuring the roughness of a texture of a surface in contact with another [9, 10].

In dentistry, the average of roughness (Ra) has been the most used and is defined as the arrhythmic medium of all profile roughness deviations in relation to the central line.

2.1.2 Wettability

Wettability is an important property with many upcoming applications in various fields. It indicates the ability of a liquid to wet the surface of solid, suggesting hydrophobic or hydrophilic tendencies. In dentistry, these tendencies affect initial water absorption and the adhesion of oral bacteria to teeth surfaces.

To determine wettability, it is necessary to measure the contact angle, which depends on the surface energy of the material and the surface tension of the liquid, formed between the surface of a material and the line tangent to the curved surface of a liquid [11].

If the contact angle formed is lower than 90°, the liquid partially wets the surface of the bare solid, meaning that the surface has good wettability properties and therefore hydrophilic.

If the contact angle formed is higher than 90°, the surface has poor wettability properties and therefore hydrophobic.

This is a simple method to gauge the wettability of a sample as well as their hydrophobic/hydrophilic tendencies [12].

The measurement of the contact angle to determine the wettability of a sample is carried out with a goniometer, an instrument that allows for a precise angular position.

2.1.3 Determination of the quantity of the element fluorine

Of all chemical elements, fluorine is the most electronegative; therefore, it is never found on earth in elemental form. Combined chemically in forms of fluoride, fluorine occupies the 17th place in order of frequency of appearance of the elements and represents around 0.06–0.09% of the Earth crust [10].

Fluoride has scientifically shown efficacy in fighting and preventing dental caries, and is widely used in most parts of the world, through its addition to public water supplies, salt, gels, topical mouthwash solutions, fluoride varnishes, toothpastes, and restorative materials [13].

During the World Health Assembly in 2007, a resolution was approved in which universal access to fluoride for caries prevention should be a basic right to human health. There are three basic methods of administering fluoride for caries prevention:

1. Those based in communities (fluorated water, salt, and milk)

2. Those administered by professionals (fluorated gels and varnishes)

3. Those self-administered (toothpastes and mouthwash) [10]

Four mechanisms of action of fluoride have been mentioned:

1. Fluoride inhibits demineralization since fluorapatite crystals, formed by reaction with enamel apatite crystals, are more resistant to acidic attacks compared with HAP crystals.

2. Fluoride enhances remineralization by accelerating the development of fluorapatite crystals by combining calcium and phosphate ions.

3. Fluoride hinders cariogenic bacteria from producing phosphonel pyruvate (PEP), which is a key intermediate of the glycolytic pathway in acid-producing bacteria.

4. In addition, F- is retained in dental hard tissue, oral mucosa, and dental plaque to decrease demineralization and enhance remineralization [14].

Fluoride remains the gold standard for stopping caries lesions with multiple systematic reviews confirming the role of fluoride products in preventing dental caries. However, F- alone does not provide a complete solution in the remineralization process, as the formation of fluoride deposits depends on the available calcium and phosphate ions found in saliva. Therefore, to increase the potential for fluoride prevention, these necessary ions have been added to formulations to increase their retention in an oral environment [15].

3. Physical properties of tooth enamel

3.1 Hardness

The hardness of the adamantine tissue refers to the resistance of the dental enamel surface to wear, scratching, or suffering certain deformation caused by the application of external pressures. It decreases from the free surface to the dentin-enamel junction, which is directly related to the degree of mineralization [1].

3.2 Elasticity

The adamantine fabric has little elasticity due to the minimum percentage of water and organic matter that it has in its composition. It is a fragile tissue with a risk of macro- and microfractures. The elasticity is greater in the area of the neck and sheath of the prisms [3].

3.3 Color and transparency

The color of dental enamel depends directly on the underlying tissues, mainly dentin, presenting a yellowish color in areas where the thickness of the enamel is less and grayish white in areas where it is thicker. It is characterized by being translucent, which is proportional to the degree of enamel mineralization; the higher the mineralization, the higher the translucency [4].

3.4 Permeability

In dental enamel, it is extremely scarce. Enamel can act as a semi-permeable membrane, allowing the diffusion of water and some ions present in the oral environment [3].

3.5 Radiopacity

Considered tooth enamel as one of the most radiopaque elements of the human body due to its high mineralization. It appears white on dental X-rays [4].

4. Chemical composition

4.1 Inorganic matrix

It is made up of calcium mineral salts, basically phosphate and carbonate, giving rise to a crystallization process that transforms the mineral mass into hydroxyapatite crystals. Mineral salt crystals are more voluminous in dental enamel, with a length of 100–1000 nm, a width of 30–70 nm, and a height of 10–40 nm [3].

They present a hexagonal morphology when they have been sectioned perpendicularly and a rectangular morphology when they are sectioned parallel to the longitudinal axis of the crystal [3]. Within the crystal are the basic units of ionic association of mineral salts called cells or unit cells, which, being associated according to the crystal, have a chemical and crystallographic composition, with calcium ions in their vertices and in the center OH^- bond [3].

4.2 Organic matrix

The main component of the organic matrix of the adamantine substance is of a protein nature, and these proteins are present in different stages of formation and maturation of the adamantine substance. There are three main proteins in the developing enamel stage: amelogenin, ameloblastin, and enamelin, which have therefore been called enamel proteins. Amelogenin is by far the most abundant protein component in the developing enamel layer, contributing more than 90% of its total volume, progressively decreasing as enamel maturity increases, and they are also called immature enamel proteins and are located between the crystals of mineral salts [5].

The ameloblastins represent 5% of the organic component and are located in the most superficial layers of the enamel and the periphery of the crystals, while the enamelanins represent 2–3%, located in the periphery of the crystals. Other proteins that play an important role are ruftelin, which is located in the dentin enamel junction zone at the beginning of the enamel formation process, representing 1–2% of the organic matrix finally within the most representative proteins found, the protein whose function is associated with the transport of calcium from the intracellular to the extracellular medium, which bears the name parvalbumin [5].

Both the timing of the stage and the degree of protein removal affect the composition of the organic matrix. Therefore, protein-mineral interactions change in enamel formation and regulate the structural organization, phase, and mineral composition, as well as crystal growth [6].

4.3 Water

It is located on the surface of the adamantine crystal, constituting to the so-called adsorbed water layer or hydration layer, the percentage of which is scarce and progressively decreases with age [5]. The unique physicochemical properties of enamel are due to its high hydroxyapatite content, the parallel arrangement of individual elongated apatite crystals in enamel prisms, and the interwoven alignment of perpendicular prisms. Together, these characteristics result in a biomaterial of great hardness and physical resilience [6].

5. Influence of the environment on the structure of enamel

Dental caries is an infectious disease that causes local destruction of the hard tissues of the tooth and is associated with diet, the accumulation of microorganisms, and the conditions of the saliva. The development of a clinically visible caries lesion is a consequence of the interaction of various factors in the oral cavity and dental tissues [8]. Carbohydrate fermentation by dental plaque bacteria leads to the formation of various inorganic acids, causing a decrease in pH. When the pH of the oral cavity reaches a critical value of 5.5, an undersaturation of $Ca + ^2$ and PO_4^{-3} occur, and ions are produced. The tendency is, therefore, the loss of ions from the teeth with the environment, which is called demineralization. This can lead to carious lesions. When the pH becomes higher than 5.5 through the buffering action of saliva, a supersaturation of Ca^{+2} and PO_4^{-3} occurs in the medium. In this situation, the tendency is to incorporate the ions into the tooth, and this phenomenon is known as remineralization [9].

There is a constant ion exchange between dental tissues and the environment, always seeking balance. Studies have shown that the use of fluorides causes a decrease in caries. A series of investigations have shown the importance of fluorides in demineralization and remineralization, and in controlling the appearance of caries, when fluoride is constantly present in the oral environment [10].

It is worth mentioning that saliva and its mucous components keep teeth moist and coated to help preserve them under the presence of calcium and phosphorus ions, thus protecting enamel from dissolution by acids. Saliva has organic and inorganic constituents. One litter of human saliva consists of 994 g of water, 1 g of suspended solids, and 5 g of dissolved substances, of which 2 g are organic matter and 3 g is inorganic matter. Sodium and potassium ions are the most abundant inorganic constituents in saliva. Sodium-ion and chloride ion concentrations increase with salivary flow rate. Among the inorganic constituents of saliva are the following: [11].

- Sodium: 14.8 mg/L
- Potassium: 22.1 mg/L
- Calcium: 3.1 mg/L
- Magnesium: 0.6 mg/L
- Chloride: 10.0 mg/L
- Phosphorus: 193.0 mg/L
- Sulfur: 149.0 mg/L

Among the organic components of saliva are glucose (200 mg/L), cholesterol (80 mg/L), creatine (10 mg/L), urea (200 mg/L), uric acid (15 mg/L). L), and other components of the parotid gland [11].

Remineralization is the natural process of carious lesion repair. Remineralization has been known for many years now. However, it is only recently that the importance of the remineralization process has been accepted as a valid therapeutic option for caries treatment. Topically administering fluoride in various forms and vehicles has shown to reduce notoriously the prevalence and incidence of dental caries. The remineralization process can be difficult if the bacterial load is high or if the salivary components are low, which is why there is a need to improve this process and apply this knowledge in a clinical setting [12].

6. Demineralization and remineralization

Demineralization and remineralization are dental caries processes that are often described as the only physical-chemical event. Although this allows for an easier understanding, description, and mechanism of the process of this disease, dental caries is much more complex. Dental caries is a multifactorial disease that includes factors such as pathogenic bacteria, salivary proteins, enzymes, ions, and fermentation of food sources (carbohydrates). This leads to the formation of biofilm, which can compromise enamel integrity, and the caries process occurs along the interface between the dental biofilm and the enamel surface [13].

7. μ-Raman

μ-Raman spectroscopy is a technique used to identify and characterize in a non-destructive and non-contact way, the chemical composition of organic and inorganic compounds by identifying functional groups without destroying the samples and without carrying out a special preparation. This technique can be used in any state of aggregation of matter [14]. With the help of the optical microscope associated with the equipment, it is possible to identify the isolated particles of the order of the micron. There are reports of the application of this technique to identify the crystallinity of human dental enamel [15].

Obtaining the spectrum is carried out using a laser that generates the beam of light incident on the sample, which is focused on a conventional optical objective in the area of interest. The same target absorbs Rayleigh (elastic) and Raman (inelastic) scattered photons. The radiation is broken down by a monochromator, analyzed using a photomultiplier, and recorded with a recording module. When a polyatomic structure is illuminated by a laser beam (monochromatic radiation of the visible spectrum), several phenomena are observed: the reflection of light, absorption, transmission, and scattering of photons. In the case of inelastic scattering, the scattered photons obtain an additive energy thanks to the energy exchange between the incident photons and the quantized energy levels of the polyatomic structure. The mechanism of this phenomenon is as follows: As a result of the action of the incident photons, which have higher energy than that of the vibrating state of the polyatomic structure, the irradiated material temporarily attains an unstable level and then returns to one of the allowed states, emitting a photon of higher energy than the initial photons [14]; the same wavelength that strikes the molecule will be the same in Rayleigh scattering (elastic), but not in Raman scattering (inelastic) as shown in **Figure 1a** and **b**.

μ-Raman spectroscopy is widely used for chemical analysis of biological and synthetic materials, in the same way it has innovated within medicine since recent studies have obtained the spectra of different human tissues, DNA, white blood cells

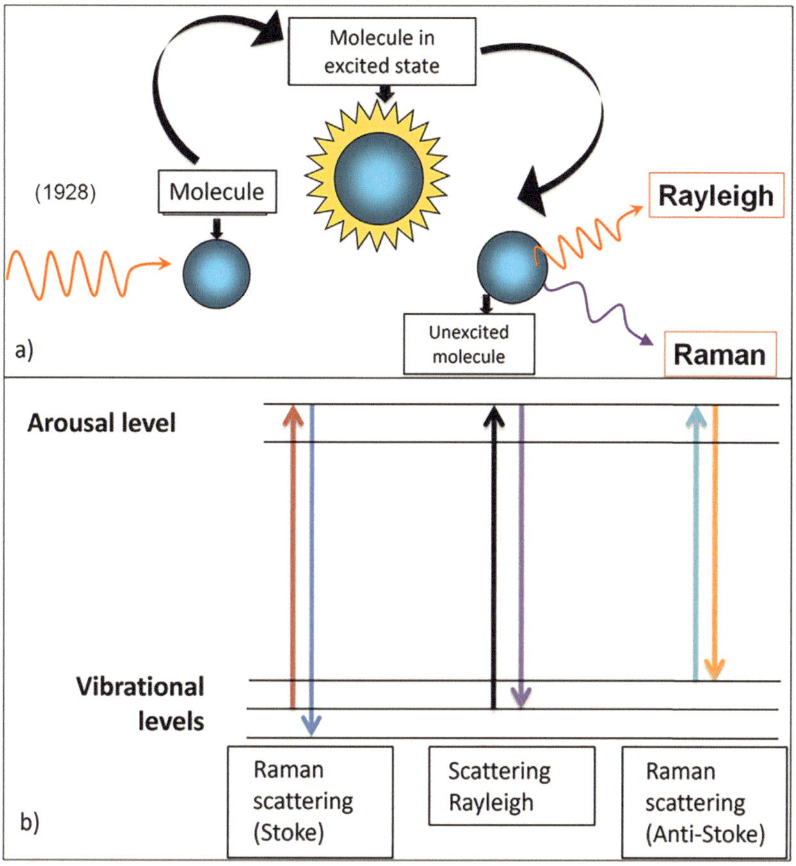

Figure 1.
(a) Representative diagram of the biphotonic effect on a molecule and the two resulting dispersions (elastic and inelastic), and (b) diagram of the energy levels. Direct source.

(leukocytes), this *in vitro* and more recently *in vivo* in studies of premalignant lesions in the chest and atherosclerotic plaques [16, 17].

7.1 Raman spectroscopic investigations

Malignant or diseased tissue is known to provoke cellular and biochemical changes in an organism, also known as tumor markers, and these changes also alter vibrational spectra. The spectrum of healthy and diseased tissue can be compared using Raman spectroscopy, which is highly specific and sensitive compared with other spectroscopic analysis in the biomedical field. This is not limited to diagnosing external tumors, but by employing miniature fiber optical probes, the technique can be used to diagnose pathologies in the oral cavity, gastrointestinal tract, and brain and ocular tissue. Other advantages of Raman spectroscopy are that the technique is non-invasive, non-destructive, uncomplicated, and the results obtained can be replicated and require minimal or no sample preparation. Applications are not only limited to soft tissues as extended research has been carried out to characterize and diagnose hard tissue pathologies such as normal and diseased bones and teeth tissues.

8. Applications of Raman spectroscopy

In the beginning, the major drawbacks of Raman spectroscopy were the fluorescence from biological tissues and the lack of sensitive instruments. Recent advances have made this technique popular due to its advantages over IR spectroscopy, such as its minimal water interference and sample preparation. IR spectroscopy requires the sectioning of samples into a specific size, which can lead to spectral changes in the chemical characterization, and this is not necessary in Raman spectroscopy where the samples can be tested without modifying their state. Modifications in Raman spectroscopy imaging have been made to characterize specific areas of interest in the medical and dental field [3]; all aimed to determined specific peaks in the spectrum, which are used as fingerprints to distinguish diseased tissue from healthy tissue.

Raman spectroscopy can also be applied to the study of milk composition as an easy, non-destructive, and fast method to determine the quality of the nutritional components found in dairy products, such as proteins, fats, fatty acids (Fas), and lactose. This makes it easy to provide consumers reliable nutritional information. [17].

Raman spectroscopy is a vibrational spectroscopy with a number of useful properties (nondestructive, no-contact, high molecular-specificity, and robustness) that make it particularly suited for PAT applications in which molecular information (composition and variance) is required. There are important applications of Raman analysis in the production of bio-pharmaceuticals, such as the characterization of raw materials, cell culture media preparation, real-time bioprocess monitoring, and the analysis of the macromolecule product,, the manufacturing and analysis of biopharmaceuticals, for example, in the biopharmaceutical production process such as Raw Materials, Cell Culture Media, Bioprocess, Protein Product, formulated medicine [18].

Raman analysis has proven to be a useful tool for the examination of synthetic and biological materials, in the medical field to study hard tissue, DNA, and white blood cells. The technique has been commonly used as a result of the development of lasers and CCD detectors, which make spectrum analysis even faster.. More recently, MRS has also been used *in vivo* to study pre-malignant lesions in breast and atherosclerotic plaques [19, 20].

There are physical characterizations of dental enamel such as microhardness, roughness, wettability, zeta potential, compressive, and flexural strength, which provide important data, but chemically, Raman is the test that complements this chemical characterization by identifying the functional groups.

9. Raman on dental enamel

In 1971, the Raman technique began to be used to identify the functional groups in mineral compounds and it was not until 1993 that it began to be used in dental research, informing about the fluorescence problems of biological materials. Tsuda [6] mentions that CaF_2 can be identified in a shift of 322 cm^{-1} in dental enamel when it has a loss of mineralization and different concentrations of fluorides are applied to the sample to reduce mineral loss; however in this investigation, the spectra were acquired in an interval of 400–4000 cm^{-1}, having this as a limitation for the detection of CaF_2. Tsuda [2] in another article mentions the use of pure hydroxyapatite crystals and dental enamel to determine the orientation of the prisms longitudinally and transversally [21].

In dentistry, natural apatites are studied, with calcium phosphates, which by adding an OH^- group form hydroxyapatite (HA), an F- form fluorapatite (FA), a Cl^- and form chlorapatite (CA); the first two are found in bone, dentin, and enamel and thanks to this characterization technique, we can observe each of these groups expressed in the bands. [18, 19].

When talking about functional groups of dental enamel, the hydroxyl groups should be located:

9.1 Hydroxyl group on Raman spectroscopy

When the concentration is increased the intensity of the band decreases, this is due to the hydroxyl stretching vibration. Additional broader bands appear at lower frequencies 3580–3200 cm $^{-1.}$ The appearance of these bands is due to intermolecular bonding, which also increases when the concentrations rise. The precision of the O-H band is dependent on the strength of the hydrogen bond. In some samples, intramolecular hydrogen bonding may occur, the resulting hydroxyl group band which appears at 3590–3400 cm $^{-1}$ being sharp and unaffected by concentration changes [22].

For solids, liquids, and concentrated solutions, a broad band is normally observed at about 3300 cm^{-1}. Overtone bands of carbonyl stretching vibrations also occur in the region 3600–3200 cm −1 but are, of course, of weak intensity. Bands due to N-H stretching vibrations may also cause confusion. However, these bands are normally sharper than those due to intermolecular hydrogen-bonded o-H groups. Particularly on dental enamel, the hydroxyl group in Raman spectra appears in the vibration range 3510–3650 cm^{-1}, when speaking of it in a clinical context [23]. When the band is observed in this vibration range, it can be translated that when the tooth is in the oral cavity it is healthy, and when the intensity of the band of this group is observed diminished, we can infer that there was an acid attack on the dental enamel that What is clinically weakening it is known as incipient caries lesion, and if that enamel is not remineralized, dental caries can appear.

9.2 Organic phosphate compounds on Raman spectroscopy

9.2.1 P-O-C vibrations

For aliphatic compound, the asymmetric stretching vibration of the P-O-C group gives a very strong broad, normally found in the region 1050–970 cm^{-1}. In the case of pentavalent and trivalent methoxy compounds, this band is sharp and strong, occurring at 1090–1010 cm^{-1}. In general, the band occurs at lower frequencies than that for the trivalent compound due to the asymmetric stretching vibration of the P-O-C group of pentavalent phosphate.

9.2.2 P=O vibrations

The band is strong due to the stretching vibration of the P=O group and in the region of 1350–1150 cm^{-1}. Taking into account the size of the phosphate atom, the frequency of the P=O stretching vibration is almost independent from the type of compound in which the group occurs and from the size of the substituents. However, it is governed by the number of electronegative substituents directly bonded to it, as well as its sensitivity to the association effects [23].

Particularly in dental enamel, the spectra of the Raman phosphate group appear in the vibration range 1100–900 cm^{-1}, where we can locate them.

Clinically, the presence of the high intensity of this band speaks of a good saturation of the dental enamel. In contrast, a weak band, slightly outside the range, speaks of a loss in the mineralization of the dental enamel, also known as incipient caries lesion or loss of continuity of dental enamel and as consequence of cavitation, that is, caries.

Figure 2 shows the characteristic bands of synthetic HA where in the vibration range 1100–900 cm^{-1} we can identify the group PO_4^{3-} (955.9 cm^{-1}) in its "strong" expression and in the vibration range 3510–3650 cm^{-1} is the OH^- group (3567.1 cm^{-1}).

A cluster of minerals containing phosphate is known as apatite, and Ha is known as calcium (Ca) apatite in mineral form and is the phosphate that most resembles the bone phosphate complex. Different salts of calcium phosphate (CaP) are summarized in **Table 1** [24].

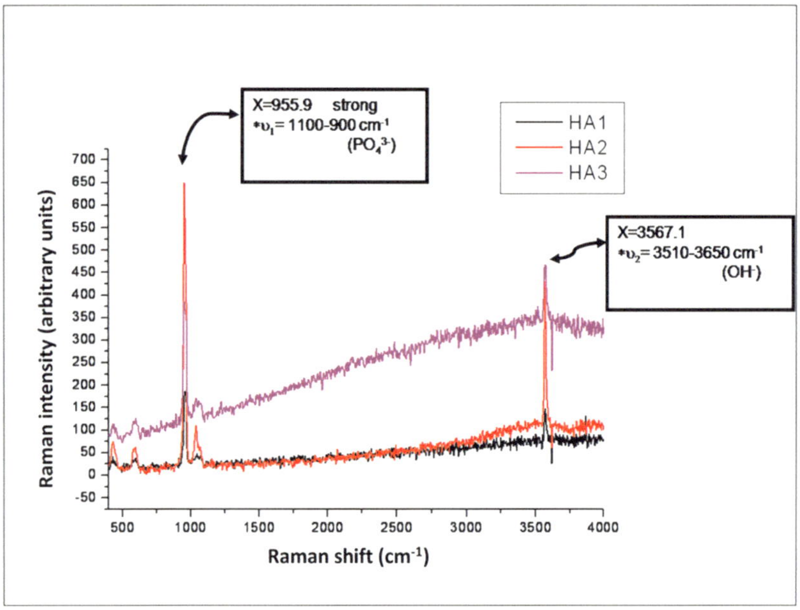

Figure 2.
Spectra of synthetic hydroxyapatite (HA) with the characteristic bands of the PO_4^{3-} group and the OH^- group. Direct source.

Name	Symbols	Chemical Formula	Ca/P ratio
Tetracalcium phosphate	TTCP	$Ca_4\ (PO_4)_2O$	2.0
Hydroxyapatite	HA or HAp	$Ca_4\ (PO_4)_6\ (OH)_2$	1.67
Fluoroapatite	FA or FAp	$Ca_4\ (PO_4)_6F_2$	1.67
Oxyapatite	OA or OAp	$Ca_4\ (PO_4)_6O$	1.67
b Tricalcium phosphate	b-TCP	$Ca_3\ (PO_4)_2$	1.5
a Tricalcium phosphate	a-TCP	$Ca_3\ (PO_4)_2$	1.5
Octacalcium phosphate	OCP	$Ca_8\ (HPO_4)_2\ (PO_4)_4\ 5H_20$	

Table 1.
Different salts of calcium phosphate.

10. Conclusions

Currently, in the implementation of different biomaterials in tissue engineering, the characterization with μ-Raman has been very useful since it does not structurally modify the sample, whether of synthetic or biological origin, and it can also be measured in any state of aggregation.

In Dentistry, it has become an excellent chemical characterization of enamel and dentin with various treatments used for mineral recovery, which is clinically known as remineralization.

Hydroxyapatite is a biological material of great interest within the apatite family, as it is a component of bones and teeth, by using m-Raman it is possible to identify the characteristic vibrational modes of the PO_4^{3-} group and of the OH^- group since the sample is not modified in its structure.

In studies where tooth enamel is demineralized or re-mineralized, characterized by means of m-Raman spectroscopy it can be concluded whether fluorinated compounds are useful or not.

Acknowledgements

This work was supported by the DGAPA-PAPIIT IA-200421 project.

Conflict of interest

The author declares no conflict of interest.

Author details

Cecilia Carlota Barrera-Ortega[1*], America Rosalba Vazquez Olmos[2], Roberto Isaac Sato Berrú[2] and Pineda Dominguez Karla Itzel[3]

1 Faculty of Higher Studies Iztacala, Laboratory Nano and dental Biomaterials, Speciality in Pediatric Dentistry, UNAM, Mexico

2 Institute of Applied Sciences and Technology, UNAM, Mexico

3 General Directorate of Military Education, Military Graduate School of Health, Mexico

*Address all correspondence to: cbarrera@unam.mx

References

[1] Sa Y, Feng X, Lei C, Yu Y, Jiang T, Wang Y. Evaluation of the effectiveness of micro-Raman spectroscopy in monitoring the mineral contents change of human enamel in vitro. Lasers in Medical Science. 2017;**32**(5):985-991

[2] Tsuda H, Arends J. Orientational Micro-Raman Spectroscopy on Hydroxyapatite Single Crystals and Human Enamel Crystallities. Journal of Dental Research. 1994;**73**(11):1703-1710

[3] Barrera-Ortega CC, Vázquez-Olmos AR, Sato-Berrú RY, Araiza-Téllez MA. Study of demineralized dental enamel treated with different fluorinated compounds by Raman spectroscopy. Journal of Biomedical Physics and Engineering. 2020;**10**(5):635-644

[4] Shin WS, Li XF, Schwartz B, Wunder V, Baran GR. Determination of the degree of cure of dental resins using Raman and FT-Raman spectroscopy. Dental Materials. 1993;**9**:317-324

[5] Requena A, Zúniga J. Espectroscopia. Madrid: Pearson Educación; 2005

[6] Tsuda H, Arends J. Raman spectroscopy IN dental research: A short review of recent studies. Advances in Dental Research. 1997;**11**(4):539-547

[7] Bona GA, Bidlack F. Tooth enamel and its dynamic protein matrix. International Journal of Molecular Sciences. 2020;**21**:12

[8] Gutiérrez M, Reyes J. Microhardness and chemical composition of human tooth. Materials Research. 2003;**6**:367-373

[9] Visscher M, Struik KG. Optical profilornetry and its application to mechanically inaccessible surfaces part I: Principles of focus error detection. Precision Engineering. 1994;**3**:16

[10] Mullane O, Baez R, Jones S, Lennon M, Petersen P, Rugg A. Fluoride and Oral health. Community Dental Health. 2016;**33**:69-99

[11] Tiznado-Orozco GE, Reyes-Gasga J, Elefterie F, Beyens C, Maschke U, Bres EF. Wettability modification of human tooth surface by water and UV and electron-beam radiation. Materials Science and Engineering: C. 2015;**57**:133-146

[12] Navarro M, Planell JA. Nanotechnology in regenerative medicine. In: MaazouZ Y, Aparicio DYC, editors. Measuring Wettability of Biosurfaces at the Microsurfaces. s.l. ed. USA, New Jersey: Springer Science+Business Media; 2012. pp. 163-172

[13] Carvalho RB, Medeiros UV, dos Santos KT, Pacheco Filho AC. Influência de diferentes concentrações de flúor na água em indicadores epidemiológicos de saúde/doença bucal. Ciência & Saúde Coletiva. 2011;**16**:3509-3518

[14] Arifa MK, Ephraim R, Rajamani T. Recent advances in dental hard tissue remineralization: A review of literature. International Journal of Clinical Pediatric Dentistry. 2019;**12**:2

[15] Gonçalves FMC, Delbem ACB, Gomes LF, Emerenciano NG, Pessan JP, Romero GDA, et al. Effect of fluoride, casein phosphopeptide-amorphous calcium phosphate and sodium trimetaphosphate combination treatment on the remineralization of caries lesions. An in vitro study. Archives of Oral Biology. 2021;**122**:10

[16] Barrera-Ortega CC, Araiza-Tellez MA, García-Perez A. Assessment of enamel

surface microhardness with different fluorinated compounds under pH cycling conditions: An in vitro study. Journal of Clinical and Diagnostic Research. 2019;**13**(8):ZC05-ZC10

[17] He H, Sun D-W, Pu H, Chen L, Lin L. Applications of Raman spectroscopic techniques for quality and safety evaluation of milk: A review of recent developments. Critical Reviews in Food Science and Nutrition. 2019;**59**:770-793

[18] Buckley K, Ryder A. Applications of Raman spectroscopy in biopharmaceutical manufacturing: A short review. Applied Spectroscopy. 2017;**71**(6):1085-1116

[19] Gilchrist F, Santini A, Harley K, Deery C. he use of micro-Raman spectroscopy to differentiate between sound and eroded primary enamel. Journal of Paediatric Dentistry. 2007;**17**:274-280

[20] Penel G, Leroy G, Sombret B, Huvenne J, Bres E. Infrared and Raman microspectrometry study of fluor-fluor-hydroxy and hydroxy-apatite powders. Journal of Dental Matrials. 1997;**8**:271-276

[21] Kirchner M, Edwards H. Ancient and modern specimens of human teeth o Fourier transform Raman spectroscopic study. Journal of Raman Spectroscopy. 1997;**28**:171-178

[22] Chen J, Yu Z, Zhu P, Wang J, Gan Z, Wei J, et al. Effects of fluorine on the structure of fluorohydroxyapatite: A study by XRD, solid-state NMR and Raman spectroscopy. Journal of Materials Chemistry B. 2015;**3**(1):34-38

[23] Socrates G. Infrared and Raman Characteristics Group Frequencies. Tables and Charts. 3TH ed. London, United Kingdom: John Wiley & Sons, Ltd;

[24] Munir MU, Salman S, Ihsan A, Elsaman T. Synthesis, characterization, functionalization and bio-applications of hydroxyapatite nanomaterials an overview. Journal of Nanomedicine. 2022;**17**:1903-1925